Upgrading Water Treatment Plants

Upgrading Water Treatment Plants

E. G. Wagner and R. G. Pinheiro

Published on behalf of the

WORLD HEALTH ORGANIZATION

London and New York

First published 2001
by Spon Press
11 New Fetter Lane, London EC4P 4EE

Simultaneously published in the USA and Canada
by Spon Press
29 West 35th Street, New York, NY 10001

Spon Press is an imprint of the Taylor & Francis Group

The authors alone are responsible for the views expressed in this publication.

Printed and bound in Great Britain by TJ International Ltd, Padstow, Cornwall

Publisher's Note
This book has been prepared from camera-ready copy provided by the authors.

British Library Cataloguing in Publication Data
A catalogue record for this book is available from the British Library

Library of Congress Cataloging in Publication Data
A catalogue record has been requested

ISBN 0-419-26050-1 (pbk)
    0-419-26040-4

TABLE OF CONTENTS

FOREWORD

The availability of safe water, and in particular safe drinking water, has been an area of concern to the World Health Organization for many years. The World Health Organization, in co-operation with the Water Supply and Sanitation Collaborative Council (WSSCC) through its Operation and Maintenance Working Group, has been involved in promoting safe drinking water and has produced manuals, teaching guides and other information on this subject.

There are many ways of satisfying the demand for water, yet one very cost-effective way is frequently not considered by decision-makers: maximising the performance of existing water treatment plants. This is done by improving the effectiveness of the existing hydraulic structure of the plants and the efficiency of the treatment process. The cost of such improvements is modest compared with the large capital investment required to construct new plants and can result in a two to three-fold increase in the volume of production. At the same time, the treatment process can be improved, which will further increase production and improve the quality of the treated water.

This book is based on extensive field experience in upgrading a wide variety of water treatment plants throughout the world. The capacity of these plants has been increased substantially and the quality of the water produced greatly improved. This has been possible because the design of most water treatment plants in developing countries was based on concepts developed in the early 1900s, when many aspects of the treatment process were not fully understood and therefore excessive compensating allowances in volumes and flow rates were made. Furthermore, in many developing countries, much of the equipment in water treatment plants does not operate properly. In many cases the problem is aggravated by poor design, specification and installation.

The concepts presented in this book are especially applicable to treatment plants in developing countries. They are therefore directed towards more labour-intensive operation and less automation, and emphasise the use of hydraulics and gravity rather than pumps and motors so as to minimise the maintenance requirements.

Efforts have been made to simplify the design of treatment units to reduce the costs of construction, maintenance and operation. Emphasis is placed on the need for chemical, physical and microbiological analysis of the raw water to be treated, collection of data on performance, and testing on a bench, pilot and plant scale.

The procedures proposed in this book apply not only to the optimisation of existing water treatment plants but also to the design of new ones. The book will therefore be a valuable source of information for designers, engineers, consultants and operators. It will also be useful for managers of water agencies, environmental health officers and students.

ACKNOWLEDGEMENTS

The World Health Organization wishes to express its appreciation to Edmund Wagner and Renato Pinheiro, Rio de Janeiro, Brazil, who prepared the initial draft of this manuscript.

The World Health Organization also recognises the valuable contributions made by the following individuals, who provided suggestions and comments on the draft text: Ali Basaran, WHO Regional Office for the Western Pacific, Manila, Philippines; Ross Gregory, Water Research Centre, Medmenham, UK; Walter Johannes, Pretoria, South Africa; Kenneth Kerri, Office of Water Programs, California State University, Sacramento, CA, USA; Yasumoto Magara, Institute of Public Health, Tokyo, Japan; and Martin Wegelin, Swiss Federal Institute for Environmental Sciences and Technology, Dübendorf, Switzerland.

Special acknowledgement is made to Ivanildo Hespanhol and José Hueb, World Health Organization, Geneva, Switzerland, who co-ordinated the production of the book and to Gordon Stott, World Health Organization, Geneva, Switzerland, who revised the draft manuscript.

Thanks are also due to Deborah Chapman, Stephanie Dagg, Ann Morgan, Susan Skelly and Colin Smith for editorial assistance, layout and production management. As the editor of the WHO co-sponsored series of books dealing with various aspects of environmental health management, Deborah Chapman was also responsible for ensuring compatibility with other books in the series.

Special acknowledgement is also made to the Operation and Maintenance Working Group of the Water Supply and Sanitation Collaborative Council and the Ministry of Foreign Affairs of Italy, which provided financial support for the activities related to the production of this book.

Chapter 1

BASIC REQUIREMENTS FOR OPTIMISATION

1.1   Introduction

Optimising water treatment plant operation is a concept applying to all plants because some operational improvements can always be made, whether in plants equipped with sophisticated instruments to monitor and control the operation, or in those with no laboratory or appropriate equipment; and whether in plants with highly trained and competent operators, or in those with operators who have little formal training but know from experience how to carry out their routine work.

Orientation for optimisation of plant operation must therefore cover a very broad spectrum of treatment plants and the people who operate them. Generalisations are difficult because each plant and situation will require a particular combination of measures to obtain optimum performance. In the industrialised world with well-equipped laboratories and well-trained personnel much of what follows may not apply, but in many less developed countries most or all of the suggested measures can bring significant improvements.

Producing best quality finished water and working at maximum capacity begins with the decision of water department managers to improve plant operation and provide the necessary resources. This implies a profound enquiry into the best treatment for the specific raw water; the application of new knowledge probably not available to the original designer; the use of the treatment plant operational history as contained in plant records; daily operator attention and monitoring of the important indicators of optimum treatment; and a renewed dedication of the operators to understand the fundamentals of water treatment, applying them in daily work.

Improved treatment plant operation does not just happen in response to decisions of policy: it is the action of many people throughout the water department that makes it happen.

1.2   Personnel

Optimising the performance of existing water treatment plants must be done mainly by plant operators — the design and construction have long been finished, and any errors or mistakes are now cast in steel and concrete. At this point, the problems which may have been built into the plant cannot easily be corrected and for all practical purposes will not be. Plant performance is therefore in the hands of those who operate it, from managers of the water department through to the operators and general workers in the plant itself.

1.2.1 Top management

In most cases where serious efforts to optimise performance are urgently needed, the water treatment plant would probably not have all the required equipment, supplies and qualified personnel. Meeting these needs will require the authorisation and support of top management including the head of the water department, the engineers in charge of water treatment and those empowered to allocate funds and personnel.

Major improvement will be possible if the water department is willing and able to provide engineering design assistance and financial support, resulting in a plant that produces two to three times more water of much better quality, and at a much lower unit cost.

1.2.2 Senior technical staff

Those most deeply involved with plant improvement are the operators themselves, but in most plants (both large and small and especially in the developing world) these operators are rarely qualified for laboratory work, or in the use of equipment such as jar-test stirrers, turbidimeters and pH meters, or qualified to make use of information from the laboratory to improve performance. It is essential therefore to have a technically trained person to be directly responsible, to lead work and to train the operators in these aspects of effective plant operation.

The best solution would be an engineer with relevant training who would understand more fully the physics and chemistry of treatment operations, follow manufacturers' instructions for equipment operation and maintenance, and use published material on plant operation and improvement. It is valuable to have an engineer as team leader, because engineering input will be essential for design and installation of some improvement measures.

The engineer should be assisted by one or more treatment plant operators who will thus acquire practical experience in making and maintaining improvements to the plant. Operators with capacity for learning and with an interest in self-improvement should be selected. If the treatment plant has an operating laboratory, the laboratory technician should also be directly involved in optimisation work, and should be aware of requirements for ongoing maintenance.

Optimisation is a continuing process and so the involvement of operators should be as broad as possible and should certainly include the long-term supervisor. In most smaller treatment plants (less than about 2.5 l s^{-1} or

200 m^3 per day) a qualified technical person (engineer, chemist or highly skilled operator) will probably not be available and will have to come from some other sector of the water department. Whatever the situation, broad involvement encourages use of the acquired information, monitoring of plant performance, and thus a sustained improvement process.

1.2.3   Plant operators

For treatment plant operators to carry out their work properly they must:

- Be trained in the fundamentals of water treatment plant operation.
- Be trained in the operation of their specific plant.
- Understand raw water characteristics at their plant and the quality standards to be met.
- Know their own specific job, know what they are supposed to do at all times and know what information is to be collected or calculated and recorded.
- Receive proper supervision and provide it to their subordinates.

Unfortunately, in most water treatment plants throughout both developing and industrialised regions, these requirements are not all met.

Few designers have any experience of water treatment plant operation and fewer operators have any input into the conceptual or final design of the plant for which they will eventually be responsible. If they had, many problems could be eliminated or at least greatly alleviated, but this is rarely the case and probably will not be for some time to come. Operators, therefore, must do the best with the physical structure for which they are responsible.

Competent, well-informed and motivated plant operators can do many things on their own to improve the performance of existing plants. These actions are, for the most part, directly related to operation but can also include simple structural or hydraulic changes. Such activities include:

- Control of the level of the intake structure from which raw water is drawn.
- Measurement of raw water flow into the plant.
- Control of the concentration of the coagulant solution in the solution tanks.
- Control of the coagulant dilution and dosage.
- Construction of an effective diffuser.
- Location of a diffuser at the point of best application of the coagulant.
- Improving the shape and location of baffles in a hydraulic flocculation basin.
- Improving the energy input of mechanical flocculation basins.
- Removal, cleaning and replacement of filter sand and support gravel.
- Building a new support-gravel layer using inverse gradation.
- Repairing the filter bottom where it may be damaged.
- Verification of filter and backwash rates.
- Control of the dosage of chlorine and of lime.
- pH correction after the application of chlorine.
- Preventative maintenance programmes for plant equipment and structures.

With technical assistance from the water department engineers and some modest financial support, competent operators can also expect to be able to do the following:

- Modify or relocate intake structures for flexible draw-off of high quality raw water.
- Design and construct a weir or flume to measure intake of raw water accurately.
- Construct and install a diffuser and piping to apply dilute coagulant at a weir or flume.
- Redesign and modify a flow-dividing manifold, if needed to distribute between basins.
- Provide proper energy input to flocculation basins by appropriate design.
- Design and build compartments for flocculation basins which use mechanical mixing, or modified baffling for those basins which use hydraulic mixing.
- Design and build auxiliary flocculation units if they are needed.
- Design and build a perforated baffle for the entrance into the settling basin.
- Design and install a more efficient system of settled water removal.
- Verify the head loss through the filter and calculate the maximum filter rate.
- Modify filter bottom and outlet piping for a filtration rate of $400–450 \ m^3 \ m^{-2}$ per day.
- Remove rate-of-flow controllers and design and install a perforated disk into the filter outlet, to control maximum filter rate for adapting the unit to declining rate control.

It is clear from these lists that operators (on their own, or preferably with both technical and financial assistance from the water department) can make very many important operational improvements. These measures can improve treated water quality, reduce treatment costs and, in most situations, increase the production capacity of the plant.

1.3   Management structure

Optimisation of plant performance is the result of effective plant operation and maintenance which depend on good management throughout the water department, starting at the top and influencing everything that the department does.

Many water departments are organised more or less as shown in Figure 1.1. Under the treatment division is the treatment plant but, depending on the city, there may be more than one plant. A treatment plant of $200–300 \ l \ s^{-1}$, operating around the clock, should have a staff of one chief operator, one assistant chief operator, three to four operators, one or two electromechanics, four to five general workers, one laboratory technician and one administrative clerk.

The chief operator is responsible for overall plant operation, maintenance and supervision of all work. Under the chief operator there may be an

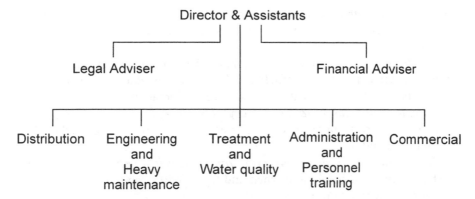

Figure 1.1 Typical structure and functions within a medium-sized water utility

assistant who substitutes for other operators in the case of illness or vacation and who is responsible for supplies of chemicals and materials, day-to-day maintenance of equipment and plant upkeep. During each shift of 8–12 hours, a designated operator is responsible for treating water at correct dosages, controlling plant flow to satisfy demand and taking note of any problems requiring the chief operator's attention. In smaller plants, the electromechanic is available during the day to work on any problem requiring specific expertise. Pumps, motors and equipment are always in need of over-haul and repair. Therefore the electromechanic does routine maintenance jobs, minor piping and hydraulic and structural work around the plant. Serious situations (including repairs which call for additional expertise) require help from the engineering division. During each shift there are one or more general workers available and during the day there can be more to do yard work and assist the electromechanics. Depending on the size of the plant, there is also a laboratory technician to do routine laboratory monitoring and testing. The clerk is responsible for producing reports, keeping inventory and other administration such as timekeeping and personnel reports.

These are the people upon whom the optimisation of treatment plant performance directly depends. Their training and motivation is a major factor towards the success of the improvement process. For complete and thorough optimisation the personnel need concerned support from top management, proper equipment, and assistance from the engineering division. A new plant can thus emerge within the structure of the old one, producing much more water of an excellent quality.

1.4   Plant maintenance

The importance of plant maintenance is obvious, yet maintenance is so poor in many cases that continued emphasis is required. The subject is large and covering it in depth is beyond the scope of this book. Thus plant maintenance is treated only in general terms, with some specific reference to pumps and motors.

Maintenance includes the use and care of plant structures and equipment, in a way that will extend their useful life and will avoid breakdowns and emergencies. General rules can be stated which cover the broad maintenance picture:

- *Provide good housekeeping — everything clean, orderly, and organised.* This is of great importance to efficient running, reflecting on the plant operation and its personnel.
- *Develop a plan of daily operation and follow it.* Everyone in the plant should know their job, and when and how to do it, so that the plant runs smoothly without problems.
- *Modify the daily plan as experience and conditions indicate.* No plan can be perfect and unforeseen conditions may require changes or alternatives. Make these arrangements an integral part of the daily plan even if they apply only in particular situations, such as excessive turbidity, so that when such situations occur the operators know what to do.
- *Follow manufacturers' recommendations for operation and maintenance of equipment.* All equipment comes with instructions. Sometimes manuals are in a foreign language and must be translated well so that important details are not lost or misinterpreted. There is a tendency always to find fault with equipment. Sometimes there may be a genuine fault, but more problems are due to misunderstanding or failing to follow instructions.
- *Establish and follow an inspection and lubrication routine for each piece of equipment.* Schedules and procedures should follow manufacturers' recommendations. To monitor implementation, these might be documented in maintenance records discussed below.
- *Keep records of maintenance and repair for each piece of equipment.* Such records, properly filed, are important for the long-term efficiency of the plant. They show which items of equipment are easy or difficult and expensive to operate and maintain, guiding further purchase or rejection. A possible example (pump record) is shown in Table 1.1.
- *Establish a plan for maintenance of the plant structures.* Routines of cleaning, painting, and repair pay dividends in long and useful service. Most of water treatment is carried out in corrosive conditions and protective coatings need to be periodically repaired. The failure to repair concrete surfaces can cause exposure of reinforcing steel with eventual structural weakening and loss. Good preventative maintenance avoids expensive waste.
- *Use photographs where possible.* Whenever the need to record a specific condition is important, a photograph can be extremely useful as an exact record of the condition at a given time.

Table 1.1   Example of a pump maintenance and repair record, illustrating typical entries

Pump:	High Service Pump No. 1

Pump data: (manufacturer, model, serial number, power rating, maximum flow rate, etc.)

Date	Maintenance and repair	By whom	Remarks
Every day	*Inspect and observe operation*	*JD*	
25-01-98	Installed new packing rings	JD	
10-02-98	Checked and greased	JD	
20-02-98	Installed new impeller	JD	Old impeller badly worn
28-02-98	Noticed hearing unusual noise	JD	Reported to Chief
10-03-98	Noise continues, so reported to Chief	JD	
11-03-98	Electromechanic inspected pump and motor	JD	Repair work scheduled

- *Maintain a well-equipped workshop with a competent electromechanic, having a reasonable stock of pipes, electrical wire and essential repair parts.* Pumps, motors and similar pieces of equipment are always wearing out, therefore a good workshop and competent mechanic are essential in maintaining the plant in good condition. Tools are sometimes lost or "borrowed" and so an investment in maintaining a reasonable stock should be made. Many people in the plant use tools and a system to account for them is advisable. An example of a suitable scheme is one where everyone who uses tools has a set of chips that are lodged in exchange for tools and retrieved when the tools are returned to the store.

1.5   Housekeeping

Housekeeping in the water treatment plant means cleanliness, orderliness and organisation. This means that the yard should not be cluttered with abandoned equipment and materials; the laboratory, warehouse and shops should be well organised and have an air of purpose about them; and electrical and hydraulic repairs should be done cleanly and securely by the appropriate personnel.

The appearance of the treatment plant, inside and outside, and the areas surrounding it give a good first indication of what may be expected. Conscientious operation is suggested by a clean and orderly plant and grounds, whilst a dirty and disorderly plant raises concern over its efficient operation and the final water quality.

Chapter 2

ASSESSING PLANT PERFORMANCE AND IDENTIFYING DEFECTS

2.1   Observing plant operation

The best way to observe plant operation is to follow the same route the water takes. Start with the raw water intake and go through the plant to the treated water reservoir. Firstly observe the operation of each unit, noting obvious problems, and begin to study the possible solutions. The next step is routine sampling to assess performance of each unit. Together with bench scale testing and pilot filter testing, an overall picture will emerge from which comprehensive improvement plans can be developed.

A study of the design of the plant is essential before making critical observations. Some plant problems are directly related to design, and there are several possible reasons for this:

- The design may not be effective for the specific raw water, particularly if good information on the raw water characteristics was not available at the time of design.
- Designs are not always suited to the level of operation and maintenance of each plant.
- Designers might not have had the best information on physical and chemical concepts in treatment plant design.
- A predesign study may have been incomplete, lacking adequate bench and pilot scale testing.

Flows, volumes, surface loadings and velocities should be checked carefully when studying the plans and specifications. Table 2.1 provides an example of such information from a plant investigation. Another important aspect of design is the ease of operation. Very few designers have operated a treatment plant and most are therefore quite unfamiliar with operation issues.

After studying the plans carefully an observer should be fully aware of the general layout and unit design performance, as well as problems that exist or are possible in the future. All treatment plants can be improved (some more than others) to produce more water of higher quality. Study and analysis of each plant can yield substantial dividends to the water agency.

Table 2.1   Unit loadings — information from a plant investigation

Grit chambers[1]		Presettling basins[2]			
Plant flow ($m^3\ s^{-1}$)	Detention time (min)	Plant flow ($m^3\ s^{-1}$)	Loadings		Detention time (min)
			($m^3\ m^{-2}$/day)	(cm/min)	
4	10.6	4	38.7	2.7	130
5	8.5	5	48.4	3.4	104
6	7.0	6	58.0	4.0	87
7	6.0	7	67.7	4.7	74
8	5.3	8	77.4	5.4	65

[1]   Volume = 2,535 m^3
[2]   Area = 8,934 m^2; volume = 31,269 m^3

2.2   Raw water intake

Considerable effort should be made to decide the best intake location, because it will provide a benefit for the life of the water treatment plant. The intake structure should be located as close as possible to the plant, but more importantly at the point in the river or reservoir where the best quality raw water (lowest turbidity and pollution) may be found.

The quality of raw water may vary greatly with depth below the surface in still waters or large rivers. In addition, the depth at which the best raw water is found will often vary during the year. It is therefore of utmost importance that the intake structure has the flexibility of drawing raw water from various depths, from just below the surface down to a point near the bottom. If the structure does not offer flexibility or is in the wrong location, consideration can be given to changes (depending on the severity of the resulting problems) (see Chapter 7).

2.3   Raw water metering

It is important that operators know the rate of raw water intake at all times, because chemical dosing is directly related to raw water flow. Unfortunately, most plants in less developed countries, and many elsewhere, have unsatisfactory measurement of raw water flow. Measuring devices that need continued maintenance or which fail easily through corrosion or wear should be avoided.

Operators tend to rely on the number and capacity of raw water pumps operating to give them the flow of water through the water treatment plant. Wear of pump impellers and surface deterioration in the transmission line means that flow will vary over time. Nevertheless, this approach can still be a

Table 2.1 Continued

Flocculation basins[3]		Final settling basins[4]			Filters[5]		
Plant flow $(m^3\ s^{-1})$	Detention time (min)	Plant flow $(m^3\ s^{-1})$	Loadings $(m^3\ m^{-2}/day)$	(cm/min)	Plant flow $(m^3\ s^{-1})$	Loadings $(m^3\ m^{-2}/day)$	(cm/min)
4	30	4	27.2	1.9	4	120	1.4
5	24	5	34.0	2.4	5	150	1.7
6	20	6	40.7	2.9	6	180	2.1
7	17	7	47.5	3.3	7	210	2.4
8	15	8	54.3	3.8	8	240	2.8

[3] Volume = 7,200 m^3
[4] Area = 12,723 m^2
[5] 16 filters, total area = 2,880 m^2

useful method if pumps are calibrated at least annually, using one of the flocculation and settling basins as a place to determine the actual volume of raw water entering the plant. For example, if the raw water pumping station has three pumps, and a standby pump, all of equal size, a settling and flocculation basin should first be drained to a known level, e.g. 2 m below the outlet elevation. The first raw water pump should then be started and pumping should be continued until the basin level has risen to the outlet level. The pumping rate is then calculated from refilled depth, basin area and pumping time. This exercise should be repeated using each pump and pump-combination in operation. This simple calibration test will give plant operators good information (typically as $l\ s^{-1}$ or m^3 per minute) for all pumping combinations with which to control proper coagulant dosages.

If a flow meter has been installed in the pipeline, calibration testing remains important and should be done each year, because meters shift out of adjustment and are not always correct.

2.4 Coagulant handling

The methods of handling chemicals in treatment plants vary widely from highly mechanised continuous systems to completely hand and batch methods. Any system can be satisfactory if designed and operated properly. What is important is that a correct dose be applied to the raw water as effectively as possible. This means certain information must be known accurately, namely:

- The dose required.
- The amount of coagulant per unit of volume in each batch.
- The amount of dilution water.
- That the dosing equipment applies the desired dose all the time.

2.4.1   Primary coagulant selection

The most effective coagulant or coagulant and polymer combination can be determined with considerable precision and economy in the laboratory. Bench scale jar testing (see Chapter 5) should be used to determine the best coagulant, combination and sequence, and the most effective and economical dosing. Unfortunately, most plants do not carry out this simple but worthwhile procedure. Relatively few water treatment plants test the chemicals and dosages routinely and continuously to search for more effective and economical processes.

2.4.2   Preparation for use

Almost all plants in less developed countries and many plants in industrialised regions use dry, solid aluminium sulphate as their primary coagulant. Usually the solid alum is put into solution in batches, preparing one or more batches whilst another batch is being applied to the raw water. This method of preparation is unreliable and unsatisfactory in most plants, because the amounts of dry alum and water are not carefully determined for each batch, which in turn is because operators do not realise the importance of maintaining an exact alum concentration. The volume of the batch tank may never have been determined exactly, or the control marks may have been lost. Most batches may be within 10 per cent of the target but much better accuracy (e.g. 1 per cent) is easily obtainable, thus efforts to control amounts of water and alum, so that concentrations are more exact, are worthwhile.

The volume of the batch tanks and the amount of dry coagulant being dissolved should be known accurately. Tank volume is easily measured and dry aluminium sulphate is most often added from sacks on which the weight is clearly, and usually accurately, marked. It is essential, however, that the operator clearly understands the importance of having an exact amount of water mixed with a certain number of sacks of coagulant to give a specific concentration. This principle of understanding as a foundation for good practice applies throughout the operations of the water treatment process.

To save space in batch tanks, alum solution prepared from solids is often too concentrated, usually in the 20–25 per cent solution range. The batch system is better designed to give a 10 per cent solution of alum (which is most economical) and makes it easier to dissolve the dry solids.

2.4.3   Application of coagulant to raw water

The most widely encountered deficiency in water treatment is in the manner of application of the coagulant to the raw water. Dilution of the coagulant down to a low concentration is very seldom done in any water treatment plant, simply because operators and plant engineers do not appreciate its importance and value. The tendency in many plants is to apply alum solution as it comes from the batch tanks. In those plants using solid alum cake this usually means a solution of about 20 per cent, and in plants supplied with liquid alum, it means almost a 50 per cent solution. In consequence it is common to observe a small, thin stream of alum solution falling into one corner of a mixing basin or onto the surface of a channel. This results in uneven dosing — a small amount of raw water receives far too much alum, while most of the raw water receives too little.

Coagulant should be applied at a concentration of around 0.5 per cent, and certainly less than 1 per cent. This provides a maximum volume of coagulant solution while maintaining a high enough concentration to avoid polymerisation and reaction with the dilution water. Plant performance observations will note insufficient dilution of the coagulant (see Chapter 7 for recommended coagulant feed systems).

Checking the system is relatively simple. Concentration of coagulant in the batch and the amount of coagulant solution being applied can be easily determined. Then, given the flow rate of raw water, the applied dose can be calculated. If the required dose is known it can be easily verified. Bench scale jar tests would indicate the proper dose but are seldom done.

The problems most commonly found in chemical feed systems are:
- Coagulant dose is not changed in response to changes of raw water flow.
- Constant head system not operating properly — the applied dose varies with the level of coagulant solution in the feed tanks.
- Chemical feed pumps out of adjustment or completely worn and performing erratically.

The consequences are that expensive coagulant is wasted, floc formation is much less than is desired, and a large proportion of the colloids pass through the filters into the treated water.

The full importance of complete and instantaneous dispersion of all the coagulant with all the raw water has been recognised quite recently. It is very difficult to attain, due simply to physical constraints, but a close approach can be made. The requirements are application at a point of high turbulence, where the velocity gradient is at least $1,000 \text{ s}^{-1}$; and dilution of the coagulant to not more than 0.5 per cent (or 5 g of solid alum per litre of water).

Rapid mixing or coagulant dispersion is very seldom attempted in the most efficient and effective way. The initial reaction of the raw water with the coagulant is extremely rapid (and is over in a fraction of a second) and therefore it is most important that all the raw water and coagulant are mixed in less than 1 second or before the initial reaction is completed.

All or most of the colloids must be exposed to a portion of the coagulant, to accomplish the destabilisation so that a floc will be formed. Positively charged metal ions (most often Al^{3+}) neutralise the negatively charged colloids and effective coagulation and agglomeration for floc formation can then occur. There are many mixing systems by which satisfactory results may be obtained. Hydraulic methods as described in Chapter 7 are recommended for less developed countries.

2.5   Flocculation systems

2.5.1   Manifold hydraulics

The transport of dosed raw water to flocculation basins is usually through an open concrete channel. Two main difficulties that can arise at this stage are ensuring an equal distribution among the flocculation basins, and avoiding excessive head loss along the route.

Manifold hydraulics must be applied wherever a pipe or channel discharges water to several points (distributing manifold), or collects it from several points (collecting manifold). In water treatment plants, distributing manifolds are often encountered in taking dosed water from the initial mixing point to a series of parallel flocculation basins; in distribution of water from flocculation basins to a series of settling basins; and in the filter backwash system, where water may be distributed to a series of transverse pipes or channels from a common header. All of these instances require application of manifold hydraulics to attain proportional (usually equal) distribution among all the points of discharge.

Distribution of water from a transport channel to a number of basins perpendicular to the channel might seem simple, and may be so when all hydraulic factors are understood and taken into account during plant design. Unfortunately for operation of most existing water treatment plants, the effective application of manifold hydraulics has been neglected in their design.

Head loss in the transport system from the rapid mix unit to the flocculation basins occurs at 90° turns where velocities are high, or at the weirs that some designs incorporate to distribute the water. Chapter 7 illustrates ways to save head loss in design and to alleviate it in existing plants. This may be important if plant production is to be increased, because head loss increases at a parabolic rate with respect to velocity and can become a major impediment.

Examples of collecting manifolds in a water treatment plant include settling basin launder systems, and stages at which a series of parallel basins discharge into a common channel. These manifolds rarely function as they should, because of improper design. When four to six basins receive water from a single channel, it is common to find that one or two basins are getting 40–50 per cent more water than others. Clearly this is bad for treatment. Overloaded basins cannot function properly and will send high turbidity water on to the filters, causing filter maintenance problems and serious impairment of treated water quality. The trouble lies in the distributing manifold and in the filter backwash system. Poor distribution of wash water from headers to take-off piping causes uneven washing and soon leads to problems in the filter bed.

2.5.2 Flocculation

The fundamental defect in most flocculation systems is that they have been designed without good information on optimum velocity gradients, flocculation time, optimum energy input or taper of energy input during flocculation. All of this is basic information on how water reacts in the flocculation basins and which bench scale jar testing can provide. Over-flocculation and under-flocculation often occur in the same basin, whether hydraulic or mechanical mixing is used.

Basins for mechanical flocculation must be divided into compartments to control the process. Short-circuiting and dead space are prevalent in basins with just one or two compartments, as most commonly found in older treatment plants. At least four compartments are needed to provide a reasonable evenness of flocculation. Direct effects of the mixing system have to be appreciated also, for example vertical rotary paddles create a higher velocity gradient at their faster-moving outer ends than near the axis of rotation.

In hydraulic systems, the velocity gradient around the ends of baffles is high while between bends it is very low. As floc particles collide and build during the process, this high and low energy input can be very detrimental and prevent optimum or even good floc formation. This results in poor settling, higher floc loads on the filters, and treated water of lesser quality.

Tapered input of energy in the flocculation process is needed to build large, settleable floc, yet few plants are designed to control tapered energy input properly. This is accomplished easily in hydraulic systems if the baffles are spaced correctly and the designer is aware of the velocity gradient to apply along the flocculation route. In mechanical mixing systems, several compartments are required with a separate agitator in each. Different energy inputs are applied in successive compartments, high at the beginning of flocculation and low at the end.

2.5.3 Horizontal flow systems

In observing a horizontal flow flocculation system, several factors should be examined:

- Appearance of the floc at the outlet.
- The number and design of compartments.
- Means of applying agitation.
- Flexibility of the agitation equipment to increase and decrease agitation.
- Time in the flocculation unit.
- Short circuiting.
- Application of a polymer, where used.

Flocculated water, as it approaches the basin outlet, should have a thick floc churning around in clouds, which are of characteristic appearance. Between the heavy floc clouds there should be cracks, which are openings between floc clouds, where the water is very clear. These may be 2–5 cm wide, perhaps several metres long, and vary continuously with new cracks appearing, closing, opening, and changing position. The floc particles themselves will vary widely in size (some being very small and some large) but a floc size of 2–3 mm should dominate.

Compartmentalisation, as shown in Chapter 7, is simply a means of making the water follow a route designed to reduce short-circuiting, aiming to maintain flocculation time close to the optimum. It should be clear in bench testing that when the flocculation time is too short, poor floc is formed which does not settle well. By contrast, when floc stays in the basin for too long it has a tendency to break up, which also hinders settling. Hydraulic, baffled flocculation systems are compartmentalised effectively by design and so there is no possibility of short-circuiting (see Figures 8.9, 8.11 and 8.12 in Chapter 8).

Once the charge on colloidal turbidity particles has been neutralised by the metallic coagulant, the particles no longer repel each other and stick together as growing floc particles. In a completely quiescent environment, contact occurs less often than if there is some agitation. Bench tests show that too much agitation shears the floc and it does not grow, but too little agitation does not provide enough opportunity for gentle contact between floc particles. There is therefore an optimum and bench testing can indicate the best floc-building environment. As floc grows, it can more readily be broken up and so the energy input must diminish along the flocculation system, i.e. agitation must be most gentle at the end of the system when the floc is largest and most easily fragmented. This applies both to the external provision of energy in mechanical mixing and to the design of baffle series in hydraulic mixing systems.

The agitation system has to maintain proper energy input across flocculation basins. Some kinds of equipment provide more even input, for better flocculation, than others (Figure 2.1). Axial flow propellers avoid the velocity-gradient variation of rotary paddles and agitate the entire basin quite evenly. Some hydraulic systems also transfer energy unevenly. At the ends of each baffle, where the water takes a 180° turn, the velocity gradient is higher than towards the midpoints of the baffles. The hydraulic system, therefore, must be an almost continuous system of turns and bends in order to be effective (see Figure 8.11 in Chapter 8).

Agitation equipment must be flexible to provide the tapered input discussed above, and to allow seasonal adjustment for changes in raw water temperature and chemical composition. Cold water requires less agitation because the floc

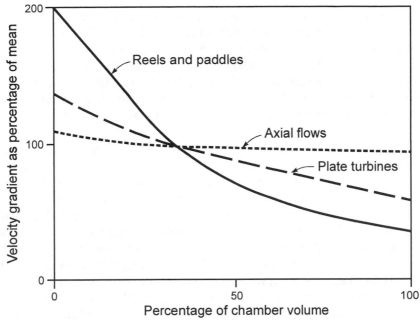

Figure 2.1 Assumed spatial distribution of velocity gradient for various impeller types

is weaker and can break up more easily. Hydraulic systems should be designed for tapered input of energy, but are less readily made flexible.

Flocculation time, short-circuiting and compartmentalisation are inter-related. Well-designed compartmentalisation keeps water in the system for close to the optimum time determined by bench tests and improves evenness of flocculation. In hydraulic systems there is no short-circuiting, but the time in the system is directly related to flow.

In some plants, heavy non-ionic polymers are used to accelerate settling. These may be used only at certain times of the year, such as in cold weather or when seasonal runoff causes problems. Other plants have continuous difficulties related to coloured water, and where it is essential to apply the polymer after the floc has formed. Bench scale testing can find the best time of application, which is often about 5 minutes after flocculation begins (see Figure 5.8 in Chapter 5).

2.6   Settling basins

After flocculation, the floc-laden water must be transported very carefully to settling basins, to avoid breaking up the floc. It might be expected that once in the basin, the floc would settle readily and relatively clear water would exit the basin; but this is not usually the case because many things happen to reduce the efficiency of settling basins.

Unless the entrance to the settling basin is well designed, energy from floc-culation mixing can carry over to form currents and short-circuiting. It is very common to find settling basins in which a large proportion of the water goes through in half or less of the design time; also there may be dead space in which the remainder stays for much longer than desired. Both of these defects reduce effectiveness of the settling basin, with poor clarification and too much floc going to the filters, where problems then follow with a poorer quality of the filtered water.

Temperature differences between the water in the basin and the water entering, cause currents and short-circuiting, again with poor settling effi-ciency and too much floc carried over to filters. If incoming water is much colder or warmer it short-circuits along the bottom or top of the basin. Sudden increases in turbidity also increase the specific weight of the flocculated water, causing it to short-circuit along the bottom of the basin.

Proper design of the entrance to the settling basin can greatly alleviate these problems and avoid some altogether. An entrance baffle (Figures 8.17 and 8.18 in Chapter 8) distributes flocculated water equally across the basin and gets it started in "plug flow" configuration, i.e. all water will travel along the basin at approximately the same velocity. Once water has entered the settling basin and distributed equally across the section, it has to be removed in the proper way. Just letting settled water drop over the outlet end of the basin is not the best method.

The settled water removal system plays an important role in obtaining water of the lowest possible turbidity for subsequent filtration. Exit velocity should be reduced to a minimum, and this requires a take-off system with the longest possible weirs or launders. The outlet weir is often only as long as the basin width, which is by no means the longest design. The result is that the velocity of the water is high, and the naturally upsweeping current carries with it a lot of floc from the basin. If the outlet weir can be doubled in length as in Figure 8.22, then water velocity over the weir can be halved; and if weir length can be tripled, quadrupled or even increased ten-fold, the overflow velocity will be correspondingly lessened. This is very desirable.

Based on sampling from hundreds of rectangular settling basins, the least turbid water is always from the middle to the third quarter of the length of the basin (see Figure 5.23 in Chapter 5). This is true in treatment plants the world over with all standards of operation, and thus weirs or launders of the settled water take-off system should extend through the final third of the basin length. Such a system reduces exit velocities and upsweeping effects, allevi-ates temperature and density currents, and takes water of the lowest turbidity for filtration. This results in longer filter runs, lower filtered water turbidity, and a less frequent need for filter maintenance.

A perforated outlet baffle is an effective and economical way to achieve the removal of the lowest turbidity raw water from the basin. The headloss designed into the baffle causes the water to exit across the end of the basin, thereby eliminating upflow currents (see Figures 8.17 and 8.18 in Chapter 8).

2.6.1 Sludge deposition patterns
The sludge pattern in settling basins provides a useful indicator of the effectiveness of mixing and flocculation, although this only applies to those basins without continuous sludge removal (i.e. those which must be drained and have sludge removed, usually two or three times a year). The profile of the sludge leaves a distinct mark on the basin wall, indicating very clearly the settling pattern. Figure 8.4 in Chapter 8 shows the ideal condition and also the pattern typical of poor flocculation.

2.7 Filters
Almost all filters used outside the industrialised nations use a single medium. This is sand of depths from 25–30 cm up to 60–70 cm, of which effective grain size varies from 0.5–0.6 mm up to 0.8–0.9 or even 1.0 mm. Although dual media filters of sand and coal are becoming more common, they are mainly used at water treatment plants in industrialised countries.

In observing and analysing filter performance, the first and most important characteristic is filtered water quality, and although turbidity has some limitations it is still the best measure of clarification for most treatment plants. With good design and operation, treatment plants should be able to produce a consistent supply of filtered water of less than 0.5 NTU. When the turbidity of filtered water is frequently more than 1.0 NTU, problems that need attention are likely.

Raw waters vary in treatability, but those producing light, weak floc need special attention in both process and hydraulic design. Preliminary bench-scale tests can identify this problem so that measures can be taken early to solve it. These measures could include use of polymers, iron as coagulant instead of alum, longer flocculation time with lower velocity gradients, lower loadings in settling basins, or special attention to velocities and turbulence. Such measures in combination can make a significant difference to the floc load that reaches the filter.

Problems relating to treatability should really be identified and analysed in the predesign phase, because the physical and process design can then incorporate whatever measures are required in response. Because few treatment plants, especially in less developed countries, are designed from bench scale, pilot scale, and plant testing, these problems are recognised only after operation begins. As a result, solutions are obviously more difficult, complex and expensive. Careful predesign investigations are most vital for water providers in regions with the least resources available for plant modification, but the reality is almost always that in such cases predesign studies are seen as a luxury rather than a necessity.

The first place to look for causes of poor-quality filtered water is not the filter itself but in pretreatment prior to filtration. If relatively high-turbidity settled water has been applied to the filter for a long period it may be subject to clogging, formation of mud balls, and possible breakthroughs in the filter beds — all of which impair filtered water quality. Causes may be found in initial dispersion and dilution of coagulant, in the flocculation and in settling.

Improvements to pretreatment are necessary before attending to the filter itself because there is no point in replacing filter sand if operation will continue with poorly pretreated water, because the filter will again deteriorate quickly. One large plant in south-east Asia has needed to change and rebuild filters constantly for 25 years. Their raw water is difficult to treat, the plant has poor initial mixing and no flocculation, and the settling basin inlet and outlet need improvement. Under such conditions, the filters will never do a good job and will always be under repair.

Although pretreatment is often the root of filter problems, there are several issues which relate to the filter itself. Properly designed filter systems are discussed in Chapter 7, and some general problems are described below.

The sand in the filter is most commonly supported by a gravel bed, below which an underdrain system removes filtered water. The same system is used for the backwash, and this underlies many filter problems — especially upset beds, breakthroughs in the beds, and poor distribution of backwash water resulting in uneven filterwash. Unsatisfactory distribution of backwash water over the filter is found in most older plants and many new ones. That part of the bed that is underwashed eventually becomes clogged and that part which is overwashed may receive backwash water at very high velocities causing an upset of the bed. Sometimes uneven backwashing occurs because of elevation differences between washwater drain troughs. The design of an effective, properly functioning backwash system requires good knowledge and application of manifold hydraulics.

Many water treatment plants (especially the older ones) were designed with an insufficient backwash rate. The bed is not expanded enough to allow a good shearing, high-velocity wash which will remove the embedded floc particles. Many systems do not provide sufficient water to give a full wash for more than 5–6 minutes, although a longer wash is sometimes necessary.

Plants may have been designed with only shallow filter boxes, saving a small amount of concrete. This produces low water depth over the filter bed and subjects the filter to negative pressures early in the filter cycle, with a small head loss. Gullets and troughs are often flooded during backwash, because they were not designed with enough capacity or grade to carry away washwater. The upper end of the drainage system floods and poor cleaning is obtained. This also occurs if part of the filter receives excessive amounts of water in uneven washing.

Air entrainment in the backwash water disrupts filter bottoms and media. If the level of the backwash header is above the level of the washwater troughs, air may collect in the header between backwash cycles. This condition occurs most often when backwash is provided by direct pumping. Air entrainment may also occur if vortices form in the elevated washwater tank as water is drained down. This problem is most pronounced in shallow tanks (see Figure 4.6 in Chapter 4).

2.8 Disinfection

Almost all treatment plants throughout the world disinfect filtered water with chlorine, although there are some, mainly in Europe, which use ozone. It is not unusual to observe one or more of the following conditions in the plant related to application of chlorine:

- Very low chlorine dose.
- Short contact time.
- Applying chlorine after lime has been added for stabilisation.
- Poorly or non-operating chlorination equipment.

To safeguard the health of consumers, the disinfection process must be complete. A low dose is effective with long contact; but contact time is more commonly short and the dosage must be high. Whatever the situation, the dose must be large enough so that the reaction will carry through to a free chlorine residual (i.e. enough to oxidise all the oxidisable material and still provide a remainder through the water distribution. Any lesser dose of chlorine will provide very precarious disinfection or none at all. In some waters where there is little or no pollution this is not critical but with highly polluted raw water, poor disinfection can be very dangerous for human health.

Chlorination equipment is produced almost exclusively in the industrialised countries. The water departments of less developed countries must import equipment and thus it is not unusual to find poorly operating chlorination equipment. Immediate reasons include poor maintenance, lack of spare parts, shortage of foreign exchange to buy the equipment and repair parts, and occasionally failure to appreciate the importance of disinfection. Chlorination equipment must be maintained in good condition to avoid the danger of a serious chlorine leak. Manufacturers' operating instructions are the best source of information regarding operation and maintenance of the equipment, and should be followed scrupulously. If the application of chlorine is erratic, the safety of the finished water is often uncertain. It is common to find the rate of chlorine withdrawal higher than the environmental temperature permits.

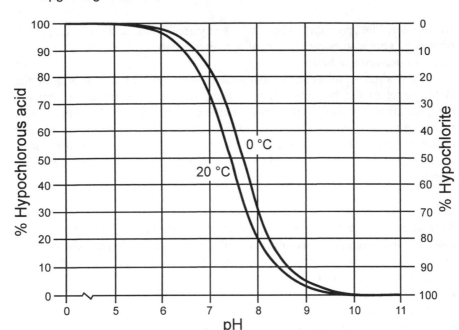

Figure 2.2 Effect of pH and temperature on the formation of hypochlorous acid for disinfection

The effectiveness of the chlorine as a sterilising agent is related to water pH. At 20 °C, the most useful oxidising and disinfecting agent (HOCl) is about 75 per cent formed at pH 7.0 but only 25 per cent formed at pH 8.0 (Figure 2.2). It is therefore essential that chlorine is applied before the pH of filtered water (normally pH 7.0 or lower) is adjusted upwards for distribution. The effectiveness of chlorine as a disinfecting agent is significantly reduced above pH 7.2–7.3, mostly wasted at pH 8, and at pH 9 or above almost no disinfection occurs.

The clear well (see Figure 8.3 in Chapter 8) should be designed so that the chlorinated water will remain in the basin for at least 30 minutes and preferably longer. This is done by baffling the tank so that the water will not short-circuit and discharge too quickly into the distribution system.

2.9   Stabilisation
Stabilisation of the treated water is very important, because water which is either corrosive or depositing can do great harm when discharged to the distribution system. Many water plants are operated without full regard for the importance of stabilisation, and the distribution system may be damaged before this problem is recognised and corrected.

Box 2.1   Checklist for review of plans and plant performance

Raw water intake
- ☑ Correct location
- ☑ Flexibility to draw water from various levels if appropriate to source

Raw water metering
- ☑ Accurate determination of flow (weir, flume or meter)
- ☑ Periodic calibration

Channels of raw water
- ☑ Calculate the maximum volume under the existing gradient

Chemical handling
- ☑ Storage capacity for each chemical such as coagulant, lime, chlorine, etc.
- ☑ Dosing capacity of each chemical (including the method, capacity and limit)

Initial mixing of coagulant and raw water
- ☑ Describe the method used, including sketches where appropriate

Application of lime for coagulation — for pH adjustment
- ☑ Describe the method used
- ☑ At what point in the treatment process is lime applied

Flocculation system
- ☑ Describe the flocculation system used
- ☑ Theoretical time of flocculation
- ☑ Volume of the system
- ☑ Type of system (hydraulic or mechanical)

Settling system
- ☑ Number of basins and their surface areas
- ☑ Surface loading
- ☑ Dimensions of unit
- ☑ Describe the entrance and exit
- ☑ Exit weir overflow rate
- ☑ Settled water turbidity.

Filter system
- ☑ Number of filters and surface area of each
- ☑ Depth and particle size of the filter media
- ☑ Support for media
- ☑ Filter bottom
- ☑ Rates of filtration and backwash
- ☑ Clean-up rate
- ☑ Filtered water turbidity
- ☑ Describe filter and wash piping
- ☑ Dimensions of drains

Disinfection
- ☑ Chlorinator size, location
- ☑ Minimum and maximum rates
- ☑ Safety precautions

Stabilisation
- ☑ Application point
- ☑ Amount applied (and flexibility)
- ☑ pH of stabilized water

Chapter 3

WATER ANALYSIS FOR TREATMENT CONTROL

Physical, chemical, biological and radioactive variables vary widely in all raw surface waters and some high concentrations may be difficult to reduce during the treatment process.

Public water supply standards vary between countries in amounts or concentrations of variables permitted and in the number for which tests must be made, for example the Environmental Protection Agency of the USA (US EPA) lists some 139 variables. Treatment for such a list is beyond the scope of this book and therefore only the most common are discussed here. When rare contaminants are suspected or confirmed, specific treatment for their removal may be needed. A list of guideline values (GV) is given in Table 3.1, derived from the World Health Organization (WHO) *Guidelines for Drinking-water Quality* (Volume 1, Recommendations, WHO, Geneva, 1993).

3.1 Physical variables

Turbidity and colour are the most common physical variables that need to be addressed during the treatment process. The levels of each of these have a large influence on the treatment to be applied and so it is important to have reliable measurements for both variables in raw water throughout the year. The necessary coagulant dose is directly related to the amount of turbidity or colour to be removed. When the levels of these variables are always low, more economical treatment can be used; if they are always high a more complex and expensive plant and process must be designed; and if they vary widely, which is common, the treatment plant must be sufficiently flexible to treat the extreme conditions as economically as possible.

The design of treatment plants must pay attention to several points when colour or turbidity of the raw water are high. The plant must have a large capacity for coagulant storage, supply and dosing; very good initial mixing of the coagulant and raw water must be achieved; and other factors must be considered relating specifically to colour and turbidity as described below.

Highly coloured waters require large amounts of coagulant to ensure coagulation and floc formation. When the colour range is 200–300 TCU (true colour units), many plants must apply 100 mg l^{-1} of coagulant to remove organic colour effectively. The floc, which is formed mainly from organic colour with little turbidity, is light, fluffy and fragile. This situation then

Table 3.1   Guideline values for micro-organisms and chemicals

Parameter	Guideline value (mg l^{-1})[1]	Consumer complaints	
		Reason for	Conc. giving rise to (mg l^{-1})
Micro-organisms			
E. coli or thermotolerant coliform bacteria	Not detectable in 100 ml sample		
Inorganics			
Aluminium	–	Deposits, discolouration	0.2
Arsenic	0.01		
Chloride	–	Taste, corrosion	250
Copper	2	Staining laundry and sanitary ware	1
Fluoride	1.5		
Hydrogen sulphide	–	Odour, taste	0.05
Iron	–	Staining laundry and sanitary ware	0.3
Lead	0.01		
Manganese	0.05	Staining laundry and sanitary ware	0.1
Mercury	0.001		
Nitrate (as NO_3^-)	50		
Selenium	0.01		
Sodium	–	Taste	200
Sulphate	–	Taste, corrosion	250
Total dissolved solids	1,000	Taste	
Hardness (as $CaCO_3$)	100	Scale deposits, scum formation (high hardness)	
		Possible corrosion (low hardness)	
Zinc	–	Appearance, taste	3
pH	≤ 8.0	Corrosion	low pH
		Taste, soapy feel	high pH
Physical parameters			
Colour	–	Appearance	15 TCU
Turbidity (median)[2]	≤ 1 NTU	Appearance	5 NTU
Turbidity (single sample)[2]	≤ 5 NTU		

[1]   Concentrations are in units of mg l^{-1} unless otherwise stated.
[2]   For effective terminal disinfection.
Source: Adapted from WHO, 1993

requires a well-controlled flocculation system to avoid any excess mixing, and careful transport of flocculated water into settling basins to avoid breaking the delicate floc. The floc going over to the filters is very gelatinous and clogs sand filters quickly.

Raw water which is low in colour but of higher turbidity (300–500 NTU or above) brings somewhat different demands. The flocculation system must be capable of handling heavy floc which may require quite high-energy flocculation because, with very low velocities or much dead space, a large amount of heavy turbidity-laden floc will settle out and cause operation problems. Transport of flocculated water must be designed to avoid settling of the heavy floc and yet avoid floc break-up. There will be large amounts of heavy sludge, so storage in the settling basin, removal and disposal must also be given special consideration.

It is clear from the above why many water treatment plants throughout the world are inefficient and difficult to operate, especially when faced with raw water of poor quality. Water departments need to collect good information constantly, because this will prove well worthwhile when they are required to provide new or more efficient production capacity.

3.2   Biological variables

The common biological variables of relevance to the treatment process are bacteria, viruses and algae. All are present in surface waters but their numbers depend on conditions in the drainage basin. Rivers in regions with large populations and industrialised areas may be highly polluted, carrying large quantities of bacteria and viruses, while most pristine streams in sparsely settled areas of the world are relatively uncontaminated and have low numbers of bacteria and viruses.

Biological variables are much more difficult and complex to monitor than most physical variables. Identification and counts of organisms demand more training and a higher level of personnel expertise than for the simple reading of turbidimeters and colorimeters. Quantification of bacteria, and in particular viruses, require more complex equipment. Reasonably large water departments should therefore have equipment and a trained microbiologist or biologist to monitor bacteria and algae. Monitoring of viruses is more difficult but probably unnecessary for most water sources.

The number of bacteria and viruses are reduced during the treatment process in close proportion to the reduction of turbidity. If treatment is reducing turbidity by 95 per cent, it may be assumed that bacterial and viral loadings are being similarly reduced. When turbidity of the filtered water is slight, subsequent sterilisation is very effective and should consistently eliminate bacteria and most viruses provided the proper dose of sterilant is applied.

From the point of view of treatment operations, by far the most important organisms are algae. These organisms can cause serious problems in treatment plant basins by the accumulation of growths on the walls, efficient clogging of filters and causing taste and odour problems. It is therefore very important to perform algal identification and counts.

Algae are present in virtually all surface waters, especially still waters (such as lakes and reservoirs) and in large slow-moving streams that are fairly clear. Raw water drawn from such situations should be monitored regularly and measures should be taken to control or eliminate the algal blooms and to prevent them from reaching the treatment plant. If control is not feasible, monitoring may enable anticipation of possible problems because algae thrive in seasonal cycles and there are times of the year when they are more or less abundant.

The water treatment process removes algae by coagulation, flocculation and settling. Nevertheless sufficient numbers may reach the filter system to cause serious problems. Growth should be controlled in the water source if possible, as well as in the treatment plant by using high shock doses of chlorine during times of algal abundance. Continuous prechlorination reduces all microbiological contaminants and keeps the treatment plant practically free of algae, but may cause a potentially worse problem of trihalomethane formation if the raw water contains the necessary organic precursors. Few water departments can monitor trihalomethanes consistently and therefore, for most, continuous prechlorination may be hazardous if the raw water carries much organic material. Periodic shock chlorination can, however, be carried out safely.

The algae of most concern in water treatment are those that cause tastes and odours, and those that clog filters. All need to be identified and controlled as far as possible upstream from the treatment plant or their presence should be given full attention when designing the treatment process. Several algal genera are cited most often as causing taste and odour problems, including *Asterionella*, *Ceratium*, *Dinobryon*, *Peridinium*, *Stephanodiscus*, *Synedra*, *Synura*, *Tabellaria* and *Uroglenopsis*. Those that are most often found to clog filter beds include *Asterionella*, *Fragillaria*, *Synedra*, *Tabellaria* and *Tribonema*.

3.3   Chemical variables and contamination
Chemical variables are by far the broadest group to be identified and monitored in relation to design and control of the treatment plant and process.

3.3.1   pH
The pH of the water indicates the degree of its acidity or alkalinity, reflecting the characteristics of the watershed or the underground rock strata through

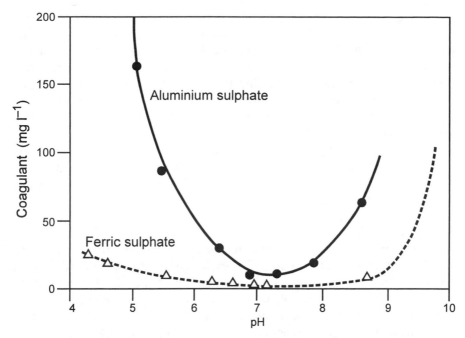

Figure 3.1 Coagulation of a water sample containing 50 mg l^{-1} kaolin, showing the dosage required for 50 per cent reduction of clay turbidity. Performance of aluminium sulphate and ferric sulphate varies differently with pH. (Modified from Packham, R.F. 1963 *Proceedings of the Society for Water Treatment and Examination,* **12**(15))

which the raw water has passed. Where limestone is predominant the waters are of high alkalinity and hardness, and have a high pH: conversely if limestone is absent the waters are lower in alkalinity, generally soft, and have a lower pH. The pH has a significant influence on the reaction of coagulants with the raw water: amounts of $Al_2(SO_4)_3$ and $Fe_2(SO_4)_3$ required to reduce turbidity to one-half of its original level vary with pH (Figure 3.1). For maximum effectiveness of alum as a coagulant, the pH range is quite narrow, while that for ferric sulphate is wide. In treatment plants using alum as the coagulant (the great majority), the optimum pH is particularly important, or coagulant is wasted.

Most treatment plants do not routinely adjust the pH for coagulation except if additional alkalinity is essential for the reaction to proceed. The coagulant dose to produce a good settleable floc is usually determined in the laboratory and through experience. The main reason for this practice is convenience — the pH may not be optimum but in the opinion of the operators it may be satisfactory. Adjusting pH with either acid or alkali requires more equipment, some expense and the attention of operators to the process. However, the final cost may be about the same.

During treatment, the pH is reduced because of the reaction between the coagulant and the alkalinity of the raw water. To avoid corrosion in the distribution system, the pH must be adjusted upwards (usually with lime) before treated water is discharged from the plant — but this can conflict with disinfection. The effectiveness of chlorine as a disinfecting agent is related to pH. If the pH is high, most of the chlorine is wasted and disinfection is not very effective, as discussed in section 2.8. The chlorine must therefore be applied before adjustment of pH for stability.

Measurements of pH should be made routinely (and recorded) for raw water, settled water, filtered water and water discharged to the distribution system. Samples should also be taken in the distribution system to monitor and pinpoint any changes.

3.3.2   Alkalinity

Alkalinity must be present in raw water for coagulation to proceed, and for a satisfactory amount of floc to form. Its origin can be natural, having dissolved from alkaline rock in the watershed, or it may have to be added because some waters are of naturally low alkalinity. The most common coagulant in treatment throughout the world is aluminium sulphate or "alum" $Al_2(SO_4)_3$. Amounts of alkali used in reaction with 1 mg l^{-1} of $Al_2(SO_4)_3$ are:

- 0.50 mg l^{-1} of natural alkalinity in the raw water as $CaCO_3$.
- 0.33 mg l^{-1} of quicklime as CaO.
- 0.39 mg l^{-1} of hydrated lime as $Ca(OH)_2$.
- 0.54 mg l^{-1} of soda ash as Na_2CO_3.

Some plants use ferric chloride ($FeCl_3$) as their primary coagulant. The amounts of alkalinity used in reaction with 1 mg l^{-1} of $FeCl_3$ are:

- 0.92 mg l^{-1} of natural alkalinity expressed as $CaCO_3$.
- 0.72 mg l^{-1} of 95 per cent hydrated lime as $Ca(OH)_2$.

Alkalinity in raw water is sufficient for the coagulation reaction to proceed in most cases (often in the region of 12.5 mg l^{-1} as $CaCO_3$, i.e. sufficient to react with 25 mg l^{-1} of alum).

The design of the treatment plant must take alkalinity into account. In addition, alkalinity must be monitored during the treatment process. If the alkalinity is too low, provision must be made for addition, whereas if it is too high, the water will be hard and softening may be necessary.

3.3.3   Iron and manganese

Iron and manganese both cause problems in water supplies. Iron is more common and occurs in silicates from igneous rocks that are widely distributed across the world, whereas manganese is found more often in metamorphic and sedimentary rocks. Iron and manganese problems occur in lakes and reservoirs where anaerobic conditions reduce Fe^{3+} and Mn^{4+} to the soluble ferrous (Fe^{2+}) and manganous (Mn^{2+}) forms. In a water body that stratifies during warm, calm conditions, soluble iron and manganese may be trapped in the bottom anaerobic water layer. When the water column becomes thoroughly mixed by wind action and changes in water temperature (i.e. turnover occurs) the soluble iron and manganese become mixed throughout the water column and cause problems for the treatment plant, due to increased levels of colour, turbidity and organic matter.

Removal of iron and manganese in most treatment plants is by oxidation of the dissolved metal ions to their insoluble state, and their subsequent removal as precipitates. Chlorine is the most widely used oxidant. Potassium permanganate ($KMnO_4$) is particularly effective in oxidising manganous compounds, but its availability and cost present difficulties for most water departments in less developed regions. Before extensive systems are designed for the removal of iron and manganese, bench-scale laboratory work and pilot investigations should be done to determine the optimum pH and time required for the oxidising reactions to occur. It may be necessary to allow a rather long holding time, of 15–30 minutes, for the oxidation to be completed. At the same time the pH must be in the alkaline range. Bench-scale tests, pilot and full plant-scale testing are discussed in Chapter 5.

3.3.4 Tastes and odours

Tastes and odours are quite common in water supplies everywhere because they are caused by a wide variety of substances, many of which readily enter water systems. Naturally occurring tastes and odours are often attributable to algae and cyanobacteria (blue-green algae). The chemical compounds which are most frequently responsible for incidence of tastes and odours include formaldehyde, phenols, refinery hydrocarbons, petrochemical wastes, naphthalene, tetralin, acetophenone, ether and other contaminants produced by petrochemical industries.

The problems are often produced when chlorine is applied to organic matter and some of the other compounds mentioned above. Where these occur it is necessary to identify first the sources and then to design remedial action. By far the best approach is to eliminate causes of taste and odour before they reach the treatment plant. This includes control of waste entering the streams above raw water intakes and control of algal blooms in water supply reservoirs. When prevention at source is not feasible then some kind of treatment remains necessary, and the two methods most often used are oxidation and adsorption. Oxidation can be achieved with chlorine, permanganate, ozone or chlorine dioxide. Bench-scale laboratory testing can be used to determine effectiveness and the likely costs. Adsorption is an effective but expensive solution, which depends on the availability of activated carbon. Treatment for taste and odour depends on its cause and the relative economy of alternative solutions. The costs of remedial treatment will always be substantial, which emphasises the importance and advantage of prevention as a first choice.

3.3.5   Sulphates and sulphides

Sulphur compounds occur widely in natural water throughout the world. Their purgative effect in excess can cause serious problems for drinking water supply. Sulphates may cause hard scaling in boilers or heat exchangers, and under reducing conditions, conversion of sulphates can supplement native sulphides to create serious odour problems.

The treatment of sulphates or sulphides is complex and expensive, such that in most cases the only solution is to tolerate the problems or to find a new source of raw water. Where there is no alternative to treatment of sulphur compounds, they can be oxidised to colloidal sulphur for removal by filtration. Some of the odour and colloids usually remain, so further treatment may be required to eliminate the problem completely.

3.3.6   Nitrates

Nitrates are a common constituent of natural waters. Important additional sources of nitrate include domestic wastes, run-off from agricultural land, and leachates from waste dumps. There is a tendency for all nitrogenous compounds to be converted to nitrates, which are therefore the most useful indicator of contamination. Nitrate removal requires costly, specialised treatment by ion exchange systems. This is beyond all but the best-equipped water providers. Because common sources of contamination by nitrogenous compounds are wastewater discharges and agricultural run-off, prevention upstream of the treatment plant is a preferable and more practical solution.

3.3.7   Hardness

Hardness is caused by divalent metal cations of which Ca^{2+} and Mg^{2+} are the most common, typically in association with HCO_3^-, SO_4^{2-}, Cl^-, NO_3^-, and SiO_3^-. In all regions where limestone is present, surface water and groundwater in contact with the rock or derived substrates have relatively high concentrations of hardness cations, particularly calcium.

There are good reasons for the removal of excessive hardness: it increases the quantity of soap needed for washing and raises costs; hard water produces scaling in units which run at high temperatures such as boilers, hot water heaters, and pipes; and unstable, hard water can form deposits which impede the flow through water distribution systems.

Hardness in the range 80–100 mg l^{-1} as $CaCO_3$ is desirable, whereas above 150 mg l^{-1} water may be considered unduly hard, and above 250 mg l^{-1} hardness is a notable problem. Where excessive hardness is recognised, the water department should analyse the circumstances thoroughly to determine the simplest and most economical treatment strategy.

Removal of hardness due to carbonates is a relatively simple process. Lime is added until the pH is in the range 9.3–10.0, which causes precipitation of calcium as $CaCO_3$. Subsequently, if required, the pH may be raised further (about 10.5) to induce precipitation of magnesium as $Mg(OH)_2$. When there is undesirable non-carbonate hardness, an additional source of alkali is added, usually soda ash (Na_2CO_3) or caustic soda (NaOH). The divalent hardness ions are replaced by sodium, although this increases the sodium content of drinking water and the increased sodium may be a problem for those who have to limit their sodium intake. After the reactions used to counter hardness are complete, the pH is too high for discharge of the water to the distribution system, and CO_2 must be applied to reduce the pH to stability.

3.3.8   Chlorine
Chlorine is used in almost all water plants for final disinfection before discharging treated water to the distribution system. Chlorination is done after filtration but before any final pH adjustment, because the proportional availability of its most effective form (HOCl) is greatest at lower pH. Design of the plant should provide for a contact period of 30 minutes to attain proper disinfection of any organic material that passes the filter, and low-turbidity filtered water reduces the possible protection of micro-organisms within particulate matter.

The dose of chlorine must be sufficient to kill organisms passing the filter, and to oxidise other material such as iron and manganese compounds. Water should leave the plant with a free chlorine residual. This verifies treated water quality and ensures protection against any contamination that might later enter the distribution system *en route* to the consumer. Colour comparator tests based on DPD (N, N–diethylparaphenylenediamine) are available for rapid and inexpensive determination of the free chlorine residual.

Figure 3.2 shows levels of chlorine and its derivatives as reactions proceed during chlorination. Initially, the chlorine is used in reaction with the most readily oxidisable materials. Next, chloro-organics and chloramines are formed, and some are subsequently destroyed, until a breakpoint is reached beyond which free chlorine residual is available. For disinfection, enough chlorine must be provided to allow this full sequence. Although chloramines have some value as a disinfectant, free chlorine is about 25 times more effective.

3.4   Laboratory analysis
Water departments should maintain at least a basic set of equipment and reagents which can be used by appropriately-trained personnel to measure accurately the physical, chemical and biological characteristics of raw and

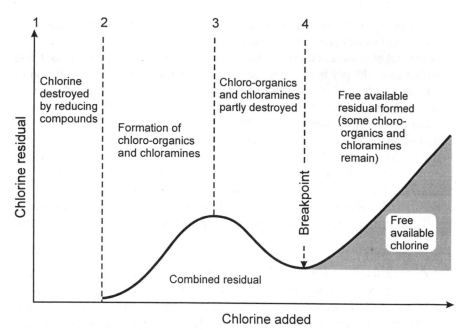

Figure 3.2 Formation of chlorine compounds with increasing addition of chlorine. A free available residual should be produced for effective disinfection during distribution.

treated water. Complete and accurate information is an essential element of water quality monitoring and of the design of treatment plant improvements. Each water department must determine its own needs, according to its resources, and then organise its laboratory accordingly.

The standard guide for water analysis is *Standard Methods for the Examination of Water and Wastewater* (19th edition) published jointly by the American Public Health Association, American Water Works Association and Water Pollution Control Federation (1992). This manual which has been used continuously since the first edition in 1905, and it is still being updated periodically. Other methodological handbooks include *Water Quality Monitoring* edited by J. Bartram and R. Ballance and published in this book series by E & FN Spon in 1996 and the *GEMS/Water Operational Guide*, produced and published by the World Health Organization (WHO) in Geneva in 1992. Most modern guidebooks now recommend the use of analytical methods that have been standardised and published by the International Organization for Standardization (ISO) in Geneva.

3.4.1 Laboratory equipment and instruments

In larger cities, it is usual to have a central laboratory that carries out a variety of analyses for the entire system. This may be equipped with sophisticated instruments and, depending on the city or the region, a large number of staff. In some places the central laboratory may be at, or near, a treatment plant. However, the functions of the treatment plant laboratory are different and it should be equipped specifically for the purpose of monitoring and controlling treatment operations. A basic list of equipment needed for analysis at a water treatment plant is given in Box 3.1.

Glassware is commonly used where cost and supply are not critical factors, but for most treatment plant laboratories (especially in less developed countries) plastic has several important advantages over glassware, such as:

- Modern plastics are hard and durable. They do scratch and mark more easily than glass, but are harder to break and will give years of satisfactory service with some care.
- Plastic laboratory ware is accurate for the requirements of common treatment analyses.
- Plastic ware that withstands the heat of sterilisation can be readily obtained.
- Plastic materials are light and transportable, which is important for field-work in rugged places.

3.4.2 Physical and chemical tests

The workload of testing and monitoring in water treatment plants needs careful scheduling, reporting and storage of records. Guidelines relating to some of the fundamental analyses are given below.

In sampling raw water taken from a river, or a reservoir with a low retention time, it should be noted that conditions change during flood periods. Thus, for example, although the hardness of such waters may be quite consistent over long periods, it is usually lowered by dilution during high flows. Some soft water rivers are prone to pulses of acidity during flood events.

- *Turbidity (by nephelometric turbidimeter) and colour (by photometer).* Take raw, settled (turbidity only) and filtered water samples during each work shift if the raw water characteristics are not changing. Under changing conditions, take samples every two hours or as often as necessary to decide appropriate responses in plant operation. Turbidity and colour determinations should be made immediately and recorded.
- *Temperature (by thermometer).* Readings of raw and final water temperature should be taken and recorded during each work shift.
- *pH (by pH meter).* Measure pH of raw, settled, filtered and finished water every two hours. Record results immediately.

**Box 3.1   List of equipment and supplies for a treatment plant
laboratory**

Equipment
- Square jars for jar testing
- Jar test stirrer with direct gear drive (e.g. Phipps & Bird)
- Refrigerator, capacity 20–30 litres (a domestic model is suitable)
- Oven for drying at 30–350 °C
- Hot plate of standard size with adjustable temperature
- Magnetic stirrer with heating element
- Still for distilled water production (2–3 l per hour)
- Single-pan balance
- Analytical balance (automatic, maximum 1 kg, accuracy 0.01 g)
- Set of sand sieves, ten over the range 0.40–3.0 mm (US standard or similar)
- Two-inch portable pilot filter with appurtenances, holding tank and pumps
- Dosing pump, peristaltic with appurtenances
- DPD comparator (for Cl determination)
- Photometer
- Spectrophotometer
- Turbidimeter
- Portable pH meter (0–14)

Glassware (or plasticware)
- Wash bottles
- Economy beakers (30, 50, 100, 250, 600, 1,000 ml)
- Graduated cylinders (10, 25, 50, 100, 250, 500, 1,000 ml)
- Burettes (50, 100 ml)
- Erlenmeyer flasks (50, 100, 250, 500 ml)
- Volumetric flasks (100, 500, 1,000 ml)
- Pipettes (1, 5, 10, 25, 100 ml)
- Funnels (10–20 cm at the top)
- Reagent bottles

Chemical supplies and consumables
- Aluminium sulphate
- Lime (commercial quality)
- Hydrochloric and sulphuric acids
- Ferric chloride and ferric sulphate
- Cationic and non-ionic polymers
- Whatman filter papers

- *Alkalinity (by titration)*. If conditions are not changing, measure alkalinity every 2–4 weeks. In changing conditions, measure as often as necessary to decide appropriate plant operation, for example once each work shift.

- *Iron (by phenanthroline method or atomic absorption spectrophotometer (AAS))*. Where iron is a problem, it should be monitored continuously. If the concentration is changing then measurements must probably be made during each work shift, but under fairly stable conditions it will usually be sufficient to measure iron weekly or less often.
- *Manganese (by spectrophotometer or AAS)*. An AAS unit is beyond the means of most water departments but could be available at a central laboratory. A simple spectrophotometer is not as accurate but will satisfy most water treatment requirements. The appropriate frequency of measurement depends on conditions.
- *Total dissolved solids (TDS) (gravimetrically)*. Determine TDS each work shift when conditions are changing — weekly or less often when conditions are stable.
- *Calcium carbonate stability (applying the Langelier saturation index)*. Determination of this stability index should be made routinely during each work shift.
- *Chlorine residual (by DPD method)*. The chlorine residual should be measured every two hours, sampling at the clearwell outlet.
- *Fluoride (by electrode method)*. Determination of fluoride should be made daily, particularly in plants where fluorine is being added in some form as part of a programme to prevent dental caries in children.
- *Hardness (by EDTA (ethylenediaminetetraacetic acid) titrimetric method)*. Hardness is quite constant and therefore weekly or, in some cases, monthly determination is adequate.

Chapter 4

IMPROVING PLANTS AND THEIR OPERATION

4.1   Plant records

Good plant records can be extremely valuable for any plans to improve plant performance, or for design of upgraded units treating the same raw water. However, critical studies for such work most commonly find that records of plant performance are incomplete, inadequate or even, for practical purposes, non-existent. If records are kept, their value for monitoring the operation of the plant is often disregarded and they are archived without analysis.

To improve plant performance one of the first factors to examine is the raw water and its characteristics, and the more information that is available the better the improvements will be. The ideal situation for older plants would be reliable data gathered over at least twenty years, including a large number of maxima and minima for important indicators such as turbidity, colour, pH, temperature, alkalinity, iron and manganese. Depending on drainage area characteristics, other indicators may be important, such as phosphate, chloride, sulphate, nitrate, pesticides, heavy metals, or various contaminants from industry.

For most treatment plants, even in industrialised countries, there would be neither time nor capability in personnel or laboratory facilities to do more than the analyses that bear directly on the treatment process. In this book, therefore, the discussion is limited to the analyses that a basic laboratory for a small water treatment plant could realistically achieve. The information that should be collected and recorded is shown in Box 4.1, including data needed for specific analyses described in this and other chapters.

4.2   Raw water intake and flow

Recommendations on raw water treatment are available for performance improvement. Chapters 5, 6 and 7 describe this work which starts with intake structures and raw water metering. Location and design of the intake are important for obtaining raw water of best quality. Operations personnel cannot easily modify the intake but it is usually possible, in reservoirs at least, to take water from a variety of depths. Sampling in the reservoir can indicate the depth where algal density is lowest, turbidity and other pollutants. If the intake is defective in design or location, the water department should be

Box 4.1   Basic information which should be recorded at a water treatment plant

General reference
- Date
- Rainfall every 24 hours
- Air temperature at 12.00h each day

Raw water
- Temperature at 06.00h, 12.00h, 18.00h and 24.00h
- Turbidity and pH each shift under stable conditions or every 2 hours if changing

Settled water (each basin)
- Turbidity and pH every 2 hours
- Residual chlorine during each shift

Filtered water (each filter)
- Turbidity every 2 hours
- Record hours of filter operation, as stipulated below

Finished water (each clearwell)
- Turbidity and pH during each shift

Chemical dosages
- Coagulant and polymer doses every 2 hours
- Lime and chlorine doses each shift

Chemical consumption
- Coagulant, lime and chlorine during every shift

Filter operation (each filter)
- Hour of backwash (time of day)
- Total time out of service (minutes)
- Hours of operation (hours)
- Duration of backwash (minutes)
- Backwash rate if available ($m^3 m^{-2}$ per min)
- Final head loss in filter before wash (m)

Raw water quality
- River level (daily)
- Reservoir level (daily)
- Intake levels

Flow rates
- Raw water into the plant (m^3 per day)
- Treated water out of the plant (m^3 per day)

Chemical analysis: each shift when water is changing, or weekly when water is stable
- Alkalinity, hardness (carbonate, non-carbonate, total)
- Iron, manganese
- Others which may be important depending on the situation

informed and if analysis of the problem shows serious defects, improvements beyond the resources of plant operators can be considered.

4.2.1 Laboratory testing of raw water

Assuming that preliminary preparations have been made, the best place to start is by testing raw water to determine optimum process parameters. Chapter 5 discusses this procedure in detail, and it is recommended that the team leader and the operators should study Chapter 5 carefully before starting the testing work. Those not trained in handling laboratory equipment and materials will need some practice, for example in using pipettes and making dilutions. Nevertheless, with an experienced team leader, sufficient accuracy in jar testing can be gained reasonably quickly. Repeated tests of some variables are very important, and therefore those operators who work routinely at the treatment plant need to master such tasks.

The time required for experienced personnel to carry out complete bench scale jar testing on the raw water might be up to 10 working days. To be realistic, an inexperienced group should initially allow three to four times as long to obtain good data, in which they have confidence. The bench scale jar testing programme should seek to determine the following:

- The most effective coagulant and optimum dosage for conditions at time of testing.
- The optimum sequence of chemical dosing if applicable.
- Whether sludge recycling would be advantageous.
- Effective polymers and the dosages required.
- Optimum flocculation time, energy input (in terms of velocity gradient) and tapering of energy input through the flocculation process.
- Settling velocity distribution curves for the test trials, and the optimum for the plant.

4.3　Rehabilitation of the chemical building

Storage and housekeeping often cause problems in chemical buildings. In most treatment plants several activities share that building such as the chief operator's office, operators' bathing and changing facilities, the laboratory, shops and general storage. If housekeeping is good there are advantages in having all activities grouped together, such as the chief is able to observe and supervise closely plant operation, capacity for interchange and communication among all personnel is good, and overall administration is simplified. With poorer housekeeping however, this situation becomes very difficult, such that plant operation is disrupted and any tendency for inefficiency or laxness spreads throughout the plant.

In general, the most difficult chemical for effective handling is lime, which is corrosive, dusty and spreads easily throughout the building if not carefully handled and confined. Lime handling and preparation should be separated and isolated as much as possible to avoid contamination of other operations. Isolation can be helped by using a separate closed area or by canvas curtain walls, or by any method appropriate for local conditions which will keep lime dust away from other activities. Lime transport by closed compressed air systems is rarely used in smaller treatment plants, and is custom-designed for each installation.

Aluminium sulphate (the most usual coagulant) is not very dusty because it is normally supplied in bags of caked solid. Some plants are supplied with alum in liquid form, which is discharged directly from trucks into storage tanks. The latter method avoids any dust.

It is rare to find an old treatment plant in which handling and application of chemicals are done well and accurately, because operators often do not appreciate the importance of applying a correct dose of fully diluted coagulant, or of ensuring complete and quick dispersion. Accuracy of the process begins with dilution in the preparation tanks (day tanks), followed by feed and dilution of coagulant and finally knowledge of raw water flow to calculate coagulant amounts. Preparation tanks have to be measured so that true volumes are known. A calibration rod should be installed in each tank clearly indicating the amount of water for a specific amount of dry chemical (perhaps a certain number of bags) to provide a solution which is about 10 per cent alum or lime. If the amount of insolubles is high, for example more than 6–8 per cent, concentrated solutions leave too much Al^{3+} in the sediment which, over a period of time, can be expensive. Smaller tanks are needed for more concentrated solutions, but the difference in construction is of minor value unless space is a particular problem.

Another important preference is that the chemicals should be fed by gravity, which eliminates pumps with their associated maintenance and operation problems. The building must be high enough to give the required difference in elevation, and where the land is sloping the building can be sited uphill.

An example is shown below of calculations in chemical preparation, relating to plant flow and the required dosage:

A maximum dose of 25 mg l^{-1} is required with a maximum flow of 100 l s^{-1}.
The preparation tank is to hold 24 hours' supply of 10 per cent alum solution.
Daily flow of water = 86,400 s per day × 100 l s^{-1} = 8,640,000 litres per day.
Amount of alum = 8,640,000 litres per day × 25 mg l^{-1} = 216 kg per day.
Volume = 216 kg per day × 10 l kg^{-1} (10 per cent) = 2,160 litres (i.e. 2.16 m^3)
The required tank size could be 2.0 × 1.0 × 1.40 m = 2.80 m^3 with 0.30 m freeboard (see Chapter 7 for details, including piping and feeding).

In practice, it is important to make the tank a size that will accommodate a specific number of sacks of dry chemical. For example, if dry alum is shipped in sacks of 40 kg, the above tank could be enlarged a little to take 240 kg of dry alum or six sacks. If the sacks are 45 kg each, five sacks would provide almost the calculated amount of 220 kg. The calibration rod can then be marked with the appropriate water level for the required number of sacks or bags of chemical.

When the exact concentration of coagulant in the preparation tank is controlled, the rate of addition of raw water can also be determined and controlled. Continuing the example above (100 l s^{-1} raw water, dosed at 25 mg l^{-1}) the amount of 10 per cent alum required is 2,160 litres per day. Thus the constant head feeder is adjusted to provide a flow of 1,500 ml per min.

The prepared solution is 10 per cent, but to be most efficient the alum should be applied at 0.5 per cent. Consequently, dilution water of 19 times the volume must first be added to flow from the preparation tank. For example, the alum feed of 1,500 ml per min would require dilution with 28.5 litres per min before addition to the raw water.

Figures 8.9 and 8.10 in Chapter 8 illustrate typical details of preparation tanks, constant head feeders, dilution control system and diffusers for coagulant application.

4.4   Pretreatment units

4.4.1   Coagulant preparation and dispersion
Maintenance of stocks of coagulant, lime and other chemicals in preparation tanks is quite simple once a routine has been established and it then only requires occasional monitoring.

Applying and diluting the coagulant or lime depends on feed equipment or chemical flow control and on the amount of raw water requiring a specified dosage. Amounts of raw water, applied chemical and dilution water must be monitored more closely and the design of the system should allow easy sampling of the chemical flow. Measurements of stock solution and dilution water flow rates can be made simply by collecting the flow for a standard time. The flow of raw water should be measurable from the weir, flume or venturi readings, and the correct dosage should have been found by jar testing.

Mixing and feeding lime involves dealing with a substance that creates dust and easily pollutes the areas where it is made up. Consequently, lime storage and handling should be isolated as much as possible from other operations. Due to its tendency to react readily with dilution water, new suspensions of $Ca(OH)_2$ should be prepared daily. To keep the lime in suspension, it must flow at a relatively high velocity and the conduits must be designed for easy cleaning.

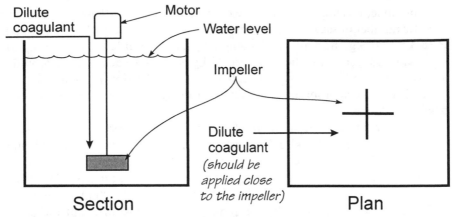

Figure 4.1 A mixing basin with mechanical agitation

4.4.2 Coagulant application
Diffusers used to apply coagulant at weirs or flumes are illustrated in Figures 8.7–8.9 in Chapter 8. Construction of a diffuser along with piping and valves is well within the capability of plant operators. In plants without a laboratory, correct dosage has to be judged from observations of floc formation and the appearance of the settled water. If mixing is mechanical, dilute coagulant must be applied as near as possible to the point of maximum agitation, close to the impeller (e.g. Figure 4.1).

4.4.3 Dividing the water among basins
In smaller plants, such as the two examples introduced in Chapters 5, 8 and 9, coagulant-dosed water would go directly to one basin; otherwise, as in the third example, water has to be divided equally among the various flocculation basins. To achieve this most simply, the exit ports from the distributing channel should be of equal size and the velocity in the channel should be constant along its length. A review of the plans and measurement of channel ports will indicate whether their dimensions are correct. If the distributing channel is in good repair, effective velocity can be calculated closely enough from flow and cross-sectional area. The channel may be tapered by design to maintain velocity with a reducing flow. Alternatively, a good solution is to calculate flow between exit ports and add filling to reduce the depth of the channel (e.g. Figure 4.2).

4.4.4 Flocculation
Operation of flocculation systems requires nothing more than to follow directions designed and constructed into the plant. Success is determined mainly by decisions in plant design. The design of the flocculation system should

Plan

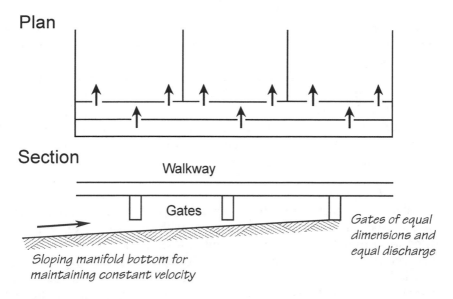

Section

Walkway

Gates

Gates of equal dimensions and equal discharge

Sloping manifold bottom for maintaining constant velocity

Figure 4.2 General layout of a dividing flow manifold

ideally be based on the results of jar testing. If this is not possible, experience has shown that a flocculation time of 25–30 minutes is often appropriate. It is important always to provide a properly tapered input of energy. The first half of the flocculation mixing cycle can be quite vigorous but the remainder should use relatively gentle mixing as the sizes of the floc particles increase. Examples in Chapters 8 and 9 discuss designs for both hydraulic and mechanical flocculation systems in detail.

The main problem in hydraulic mixing is the long baffle walls, where flow is quiescent and little mixing occurs; the only mixing occurs at the bends. Baffle walls can be filled with perpendicular struts or fins to cause agitation as the water flows between bends, as shown in the examples that use hydraulic mixing in Chapter 8. Hydraulic flocculation systems may also accumulate sludge, especially if the actual flow is less than assumed in plant design. The lower velocities encourage settling of heavy, fully formed floc particles at the end of the unit.

Mechanical flocculation requires good maintenance of motors and agitation equipment. The rotary speed of the agitators must be close to that of the design, and should provide tapered energy input by running at decreasing speeds through the course of flocculation. Problems with mechanical flocculation are usually in the compartmentalisation of the system. If the water is left to take its own route it will usually short-circuit. It is not unusual to find as much as half of the water having only 5–10 minutes flocculation time. A predesigned route should be constructed with baffles and walls, so that all of the water stays in the system a full 25–30 minutes. Chapter 7 shows that quite simple changes can be sufficient to obtain much better results in mechanical flocculation basins.

If bench-scale testing equipment is not available to find the optimum velocity gradient along the flocculation path, a safe assumption would be 30–70 s^{-1} during the first half of the cycle and around 20 s^{-1} for the remainder. The operators, with a minimum of help from the engineering department, can make the adjustments in flocculation to improve performance significantly.

4.4.5 Settling basins

Almost all rectangular settling basins experience hydraulic problems at their entrance from the flocculation system and with the process of removing settled water from the basin. The first aim is to bring water into the basin at equal velocity across its section, so that all water starts flowing down the basin at the same velocity (in plug or piston flow). Chapter 8 discusses, with examples, the calculation of dimensions for a perforated entrance baffle that introduces a head loss in the flow. Water will always take the route which incurs the lowest head loss. Because all the ports have equal head loss, the water will enter evenly across the basin. Plans and observation of the basin will show whether or not a baffle is needed. Water exiting the flocculation basin should not flow directly on the perforated baffle; therefore a blind baffle is placed in the line of flow to absorb and reduce the energy and to distribute the flow.

The perforated baffle can be designed and installed regardless of bench-scale jar testing work. In plants having the equipment for jar tests the velocity gradient in the baffle should follow that of the last portion of the flocculation system. If no jar testing has been done, experience suggests that a maximum velocity gradient of 30–35 s^{-1} can be used for water above 10 °C. Alternatively a limit of 20–25 s^{-1} can be used for colder water.

Generalisations are less reliable for basin loading than for velocity gradients and their tapering. Specific bench-scale testing is needed to determine the optimum settling velocity, otherwise it is prudent to use a settling velocity or surface loading which is conservative. Loading at 2.8 cm per min (about 40 m^3 m^{-2} per day) may be sustainable, and it is possible that the loading could be more. However, without testing there is no sure guide.

It is valuable to have good loading information, because flow through the flocculation system is often the limiting factor in plant improvements. Increasing flow rate reduces flocculation time in an existing unit, perhaps until it is ineffective. If there is severe short-circuiting the actual flocculation time may be quite short. Correction of such a defect will improve flocculation time so that loading can be increased without detriment to floc formation.

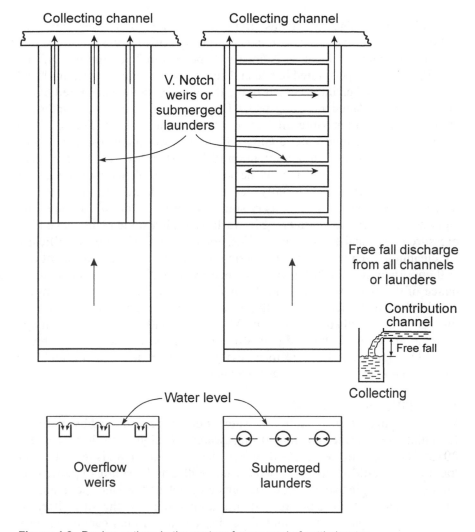

Figure 4.3 Design options in the system for removal of settled water

The most common type of settled water removal from the basin is a simple weir across the end of the basin. The volume of water per unit length is very high over this relatively short weir design, causing an upflowing current which carries a great deal of floc which could otherwise have settled. Furthermore, the least turbid water, which is usually in the middle of the basin, remains there, with the result that the best quality settled water is not withdrawn.

Figure 4.3 shows the most efficient ways of removing the settled water from the basin. The key to success is a long length of weirs or launders, unflooded at their discharge points. With this design overflow rates are very low, the upflowing currents are trivial and the best quality settled water is removed.

Financial and technical assistance will be required for the design and installation of the perforated baffle and settled water removal system. Head losses must be taken into account in the design of both the perforated baffle and submerged launders. Long weirs for removal of the settled water are simpler in design but more difficult to construct and install. The resulting improvement in settled water quality will be well worth the cost and effort.

Settling basins need further attention if the system involves sludge accumulation, where every 3–6 months the basins are each taken out of service for sludge removal. Basins are usually cleaned with hoses and high pressure jets, maybe using adjacent basins as a source of water. Portable pumps and siphons can be employed to bring in wash water, but some plants have high pressure lines installed along the basins, and these are to be used specifically for cleaning. Even in basins with continuous sludge removal equipment it is still important to drain the basin about once a year, for inspection and maintenance of the cleaning equipment. Basins that use hydraulic sludge removal should also be cleaned periodically, because some sludge always remains in the basins and should be washed out. When basins without continuous sludge removal are drained, the sludge profile line will show clearly on the basin wall. This mark can be used to assess proper operation and it should be similar to that shown in Figure 8.4, Chapter 8.

4.5   Filter rehabilitation

Where coagulant dispersion in raw water is defective, followed by marginal flocculation and inefficient settling, water of high turbidity (perhaps 20–25 NTU) enters the filtration system. This turbidity load is too much for the filter to handle for any extended period. With a high loading of particulate matter filters begin to deteriorate, become clogged, and produce poor-quality filtered water. At most plants with filtration problems, the filter itself is blamed. Although that may sometimes be the case, filters cannot perform well under excessive turbidity loads and the problem is often with prior units that are ineffective. The first step to improve filter performance is to ensure that treatment prior to filtration is of a high standard and only then should the filtration system be assessed in detail.

For plants more than 10 years old, filter rates are usually around 110–120 m^3 m^{-2} per day and backwash rates are typically less than 0.3–0.4 m per min. The low filtering rate does no harm except that the potential capacity is not fully used, but the low backwash rate will not clean the filter properly. Each wash leaves a little floc behind and the cumulative effect is a dirty, badly clogged filter. This process can be accelerated by settled water of higher turbidity.

Most older plants also have unsatisfactory distribution of wash water across the filter area, with some parts overwashed while others are underwashed. In addition to inefficient washing itself, the support gravel may be disturbed by uneven washing. Another common problem relates to the backwash water drain system. Very often the channels are too small or do not provide sufficient flow velocity to carry the wash water, which is held back and causes underwashing of that part of the filter close to the outlet.

If a filter has been overloaded with highly turbid water for long periods it will be full of mud balls, the surface will have cracked and sand will have pulled away from the walls. At this stage, the only recourse is to clean the filter thoroughly, removing and washing the sand and support gravel to examine the bottom and, if necessary, to remake the entire filter.

Improvement of filter performance can be a major operation, requiring a careful plan for overall plant improvements, management support and close collaboration of personnel. The best approach to filter rehabilitation is to take them out of service one-by-one for complete renovation. This can be a lengthy process but many older plants will have gone for years without over-haul, and a careful filter-by-filter programme need not cause great operational problems. Normally, with only one filter out of use, the others should be able to take the overload for a short time. While carrying out this work, operators gain proficiency with the renovation of each filter. The first filter is a learning experience and will take the longest to renovate. Depending on filter size, experience of personnel and the improvements required, it may take one to two months to complete work on the first filter, then half as long on the remainder with the benefit of experience.

There are sufficient points of caution in filter renovation that a detailed description of the main stages is justified here, based on experience gained with a variety of situations.

Detailed planning is a prerequisite for effective filter rehabilitation, because so many components are involved — scaffolding, passageways, wheelbarrows, hand tools, clear areas, sand-washing equipment and replacement piping, fittings, bottoms, etc. All of these should be prepared and in place before any actual filter work begins. Replacement sand and support gravel should have been screened, classified, and piled, protected from the weather. Plans of the filter design should provide details of the filter piping structure and, if changes are planned, the pipe sizes and fittings can be partially assembled ready for installation. If no changes in filtering capacity are planned, the work is greatly simplified.

Special care is necessary in placing support gravel and the sand and coal media in the filter box. First and foremost, the gravel should be as round as possible and in particular all flat stones must be removed. The appropriate levels for the layers of support gravel should be carefully marked on the filter walls. After marking, allow water to enter, to verify the consistency of levels on all sides of the filter box. When the levels have been marked all around the box, begin with placement of the bottom gravel layer.

With the first layer placed, lay a wide board on top of the gravel at one side of the filter box and from this point on, always work from such boards. Never allow personnel to step directly onto the gravel layer because this creates holes in the support layer that can cause serious problems later. When each gravel layer is complete, allow water in to the depth marked on the wall and use this to make sure the layer is level, raking to avoid high or low areas.

Once satisfied that each gravel layer is correct, begin installing the next. Never dump or drop gravel into the bed but lower it in buckets of a size that can be handled easily. This continues until all gravel layers have been placed and levels confirmed with water.

Next add the washed, cleaned, graded and dried sand, with the level also marked on the filter walls. Observing the same precautions, sand is carefully layered over the support gravel. With this reverse gradation, spaces among the top gravel layers become filled with sand. Cover the top of the gravel with about 10 cm of sand and rake carefully so that spaces will be filled. Once the sand layer has been completed to the level marked, it can be verified with water.

With the sand in place, backwash the filter at about 50 per cent flow by opening the backwash valve slowly until the backwash flows at the desired rate. This will help to clean any dirt that may have entered and will distribute sand better in the top gravel layers. After several minutes at this wash rate, increase slowly again until the sand is at 30–40 per cent expansion and continue at this rate for at least 10 minutes, after which the backwash water should be very clear. Slowly close the backwash valve, then open the filter valve and drain the filter down to a point below the sand level. Verify the sand level. The level will probably be a little low because more sand will have entered gravel spaces. If necessary, more sand may be added to reach the proper level. Repeat the backwashing as before.

When the sand level remains firm, add the coal (if this is being used), carefully levelling it in 10–15 cm layers to the required level. Verify the level with water entering slowly from the backwash. The coal should be about 3 cm above the required level, because some of the fines will be washed out and the coal compressed. Start the backwash again at about 50 per cent and then, after about 10 minutes, increase the wash rate to 75 per cent and continue until the filter is very clean. Finally, increase slowly to full backwash and check carefully for loss of coal. The fines should wash out but not the larger particles.

All dual media filters have an auxiliary air or water wash, which should be tested after all media are in place. Auxiliary washing is discussed fully later in this chapter and general approaches are summarised below.

Air washing operates with the water level in the filter at the level of the drain trough or gullet. With the water at this level, the drain valve open and all other valves closed, start the air slowly and continue for 4–5 minutes. Air should bubble up through the media and agitate the sand and coal vigorously, such that the water will become very turbid when the filter is dirty. Close the air valve slowly and start the regular backwash. The few minutes of air wash will have given the bed a vigorous scour and the water will complete this task.

For beds fitted with sweeps, the water level in the filter should be lowered to just below the level of the sweep. Start by opening the high pressure line. The sweeps will begin to rotate and gain speed rapidly. As this is happening, begin the backwash. The sweeps will be covered quickly and will rotate more slowly within the expanded bed. Sweeps should be allowed to operate until about the last minute of the backwash.

The static water wash system is also started with the water just below the top level of the jets. Once the system is operating, the backwash also begins. As with sweeps, the static water wash system operates throughout the backwash until the last minute of wash. First the auxiliary system is closed, then immediately afterwards the backwash is stopped.

In all these three systems the integrity of the air or water pressures and quantities should have been tested during construction and all jets checked for proper function. Sometimes the auxiliary wash is with a high pressure jet from a hand-held and controlled hose at the top level of the filter box.

4.6   Filter operation
When the initial coagulant dispersion, flocculation and settling has been brought up to the optimum the settled water will be of low enough turbidity that the filter can perform well. Once this pretreatment is in good order, filters can be rehabilitated to the extent necessary to put them in good operating condition — the turbidity of filtered water should be always less than 1.0 NTU. Maintenance of pretreatment standards is necessary to maintain the filter in good condition, but some procedures and points of caution apply to the filters themselves.

Operators should observe the bed during backwash cycles, noting areas in which water boils up at a higher velocity than elsewhere (Figure 4.4), which is usually evidence of a broken nozzle or a break in the filter bottom. These boils often dislocate support gravel, making a hole, which gathers sand. Eventually the filter begins to lose sand through the broken filter bottom. This problem usually becomes worse with increasing sand loss from the bed.

Good quality filter sand is expensive because of the work involved in preparing it to the correct specification. Thus losing sand through broken filter bottoms is wasteful and uneconomical. The clear well should be inspected carefully and frequently and if sand is found the problem filters

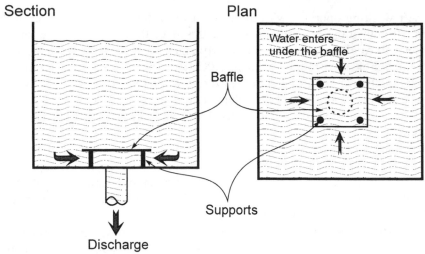

Figure 4.4 Anti-vortex baffle installed in washwater tank

should be identified and repaired. Damage to the underdrain and filter support gravel usually occurs in three ways, which are discussed below.

Most often, air slugs become trapped in the backwash water and burst in the underdrain system, causing high localised pressures and velocities. The result may be broken nozzles or pipe or concrete, which encourage movement of gravel and subsequent loss of sand. There are two common sources of slugs of air in the backwash water. If the washwater line is above the level of the water in the filter, there is a good chance that air will collect in the line between washes. The only simple remedy is to maintain pressure on this line constantly. If this is not possible then the level of the wash water line must be lowered so that it is always under pressure from water in the filter. Secondly, when the washwater tank is drawn down to a certain level a vortex forms, entraining air which goes to the filter. The solution here is merely to install anti-vortex baffles above the point of discharge (Figure 4.4).

A second cause of underdrain damage is from high-velocity backwash water. If this enters the underdrain system suddenly and violently, its high energy must be dissipated by the receiving structure (pipe or wall) which may fail if it is not sufficiently resistant. Therefore, backwash water valves should be opened very slowly and also closed slowly, to reduce the chance of sudden destructive surges.

Underdrain damage can also be caused by operation of the filter to a point of excessive head loss and thus a high negative head or vacuum in the filter bottoms, leading eventually to a collapse of some of the underdrain structure. This problem is obviously alleviated by washing the filter before head loss is too high, but in many filtration systems the operator has no indication of this.

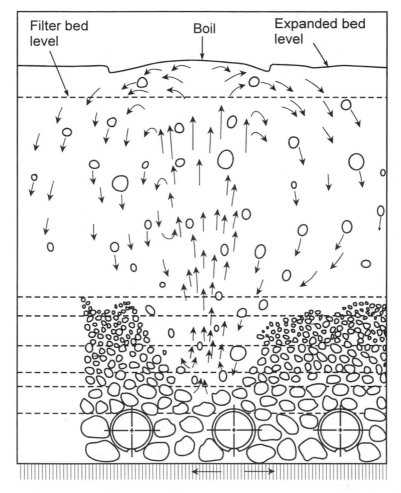

Figure 4.5 Filter disruption and short-circuiting, with bed displacement and a surface boil during backwash

It is quite simple to install piezometer tubes on the filters so that operators can readily monitor the head loss. Figure 4.5 shows the anatomy of filter disruption, and Figure 4.6 shows the hydraulics involved in air binding which brings this problem.

Experience illustrates clearly the advantages of auxiliary air or water wash: either system will maintain the filter in much better condition than just water backwashing alone. There are four ways that auxiliary washing can be accomplished. Auxiliary air agitation must be designed into the filter system and the filter bottom must also be designed to provide for air entry and movement through the filter bottom and support. Auxiliary water wash can

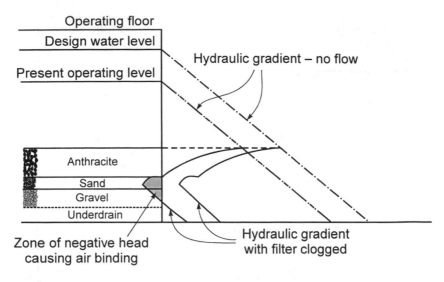

Figure 4.6 Diagram to show the cause of air binding within filter media

be applied through rotary sweeps which are installed over the filter and which operate 5–8 cm above the surface. With the filter expanded, sweeps rotate within the expanded sand (or coal in the case of dual media). As an alternative to sweeps, water washing may be achieved from a stationary grid with jet nozzles attached at intervals. This system is installed at two levels in dual media beds: the top grid is set 5–8 cm above the unexpanded surface, and the lower grid lies about 5 cm above the sand-coal interface.

The fourth method is to apply a hand-held high pressure jet throughout the wash cycle. Many older plants were designed without auxiliary air or water wash and have always operated that way. It is often beyond the means of water departments to install automated auxiliary systems in existing plants, but hand-held jets can be used at little extra cost. Throughout the backwash while the bed is fluidised, a high-pressure jet can be directed into the bed, so that the bed area is covered several times and walls are also washed down. This provides a great deal of extra agitation and therefore additional cleaning power. With proper conditions (high pressure available, applied through a good hose and nozzle designed to provide a high-velocity jet stream) filter beds can be maintained in excellent condition. This practice should be adopted in all plants designed without auxiliary washing. If water pressure is too low, a small pump can be installed specifically to increase pressure during the wash. When coal is used over sand in dual media filters the auxiliary wash is essential and can be done by this system, but it is not automatic and operators must make the effort to do the work during every wash. The general backwash procedure should be as follows:

- Close the settled water inlet valve or gate.
- Continue filtering until the water is 20–30 cm above the backwash troughs or gullet.
- Open the drain and the water will quickly reach the level of the troughs and gutter.
- Continue to filter until the water level is 15–20 cm above the unexpanded bed.
- Close the filter water valve.
- Start the auxiliary air or water wash and continue for 3–5 minutes.
- Open the backwash valve slowly until it is fully open.
- Close the auxiliary air system. With water jets, continue for 5–6 minutes during full backwash, and then shut off the auxiliary wash water valve.
- Continue the backwash until wash water is clear. Within 6–8 minutes the backwash water should have a turbidity of 5–10 NTU.
- Close the backwash valve slowly.
- Return the filter to service by opening the settled water inlet valve and filter valve.

The second step is often not followed and instead the drain valve is immediately opened after closing the inlet, which wastes a great deal of water on each wash. The filter box drains quickly and the filter returns to full service with less delay, but in most plants a reduction of filter rate for the last 30–60 minutes of the run is outweighed by the economy of water. Exceptions would include plants where raw water is abundant and reaches the plant by gravity with no expensive pumping involved and where filters are washed very infrequently, for example every 80–96 hours of operation. If filters are washed every 24–30 hours and pumping is involved, allowing filtration to continue until the level is relatively close to that of the trough and gullet is economically advantageous.

Depending on the level of the backwash overflow drain troughs and gullet, and on the backwash rate, there is always the possibility of media being washed out of the filter. To check that the backwash rate is not excessive, drain troughs and gullet should be carefully inspected for sand or coal after the wash and before putting the filter back into service.

4.7   Stabilisation

One important treatment function poorly understood by operators everywhere is the need for production and distribution of stabilised water. The distributed water should be neither corrosive (aggressive), reacting with metals to damage the system or domestic plumbing, nor depositing, leaving a calcium deposit to clog the system. Water in granitic areas and other resistant geology is soft with a naturally low pH and attacks metals with which it comes in contact. The surface and ground waters in such areas, including most of Africa, South America and large parts of Asia, may tend to be aggressive and must be stabilised before distribution.

The usual way to stabilise soft water for distribution is to add lime, thus raising the pH, hardness and alkalinity. Some lime deposits are found naturally in almost every country. The addition of coagulants such as $Al_2(SO_4)_3$, $FeCl_3$ or $Fe_2(SO_4)_3$ reduces pH during treatment and chlorine further reduces the pH if applied as a gas, hence treated water is usually aggressive. Lime or another alkali is therefore added to correct pH and stabilise the water.

The Langelier Index provides a simple means of determining whether or not the water is close to stability. This test requires some equipment and chemical analysis. It is necessary to measure the total dissolved solids (TDS), pH, calcium concentration and alkalinity for the calculation of stability as follows:

$$\text{Langelier Index (LI)} = \text{pH of the water leaving the plant} - pH_s$$

where pH_s is the "pH of saturation" which is calculated by:

$$pH_s = \text{temperature factor} + \text{TDS factor} - \text{alkalinity factor} - \text{calcium factor}$$

Temperature, total dissolved solids, alkalinity and calcium factors are given in Table 4.1, noting that alkalinity and calcium factors use the same conversion from concentration. If LI is less than zero these aggressive waters tend to dissolve $CaCO_3$, whilst LI greater than zero indicates depositing waters from which $CaCO_3$ tends to precipitate.

Measurements for a typical sample of treated soft water might be pH 7.0, TDS 90 mg l^{-1} (factor 9.76, interpolating from the values in Table 4.1), alkalinity 40 mg l^{-1} as $CaCO_3$ (factor 1.60), calcium 60 mg l^{-1} as $CaCO_3$ (factor 1.78) and temperature 20 °C (factor 2.10). For this situation, pH of saturation and Langelier Index would be given by:

$$pH_s = 2.10 + 9.76 - 1.60 - 1.78 = 8.48$$
$$\text{Langelier Index} = 7.00 - 8.48 = -1.48$$

This value of the Langelier Index suggests a highly aggressive treated water. Consequently, calcium and alkalinity must be raised, usually by the addition of lime. Each increase of 1 mg l^{-1} as $CaCO_3$ requires 0.74 mg of lime; thus if 25 mg l^{-1} of lime is added, the calcium or alkalinity will be increased by 34 mg l^{-1} as $CaCO_3$. The resulting calcium value would be 94 mg l^{-1} as $CaCO_3$ (factor 1.97) and the alkalinity would be 74 mg l^{-1} as $CaCO_3$ (factor 1.87), giving a revised pH of saturation as:

$$pH_s = 2.10 + 9.76 - 1.97 - 1.87 = 8.02$$

Table 4.1   Temperature, total dissolved solids, calcium and alkalinity factors for the Langelier Index

Temperature		Total dissolved solids		Calcium or alkalinity	
°C	Factor	mg l^{-1}	Factor	mg l^{-1} CaCO$_3$	Factor
0	2.60	0	9.70	10	1.00
4	2.50	100	9.77	15	1.18
8	2.40	200	9.83	20	1.30
12	2.30	400	9.86	25	1.40
16	2.20	800	9.89	30	1.48
20	2.10	1,000	9.90	35	1.54
26	2.00			40	1.60
30	1.90			45	1.65
				50	1.70
				60	1.78
				70	1.85
				80	1.90
				90	1.95
				100	2.00
				110	2.04
				120	2.08
				130	2.11
				140	2.15
				180	2.26

After adding 25 mg l^{-1} of lime, the actual pH will also have changed to around 8.00, and the revised Langelier Index would be:

$$\text{Langelier Index} = 8.00 - 8.02 = -0.02$$

This result is almost neutral and the water should be neither aggressive nor depositing. It should be emphasised that the Langelier Index is an approximation, but will provide a sufficiently reliable guide for protection of the distribution system and consumer interests.

One of the best direct methods of monitoring stability of the water in the distribution system is by small steel plates placed in the pipelines. The plates are carefully weighed and placed in a fitting where they are held securely. Then, after about three months, the plates are removed and reweighed. They will weigh less if aggressive water has dissolved the steel and will weigh more if deposition has occurred. If the water has remained stable there should have been no change.

Chapter 5

DETERMINING DESIGN PARAMETERS

5.1   Introduction to bench scale testing

Very few of the water treatment plants in operation today were designed for their specific raw water supply, under the variety of conditions that occur during the year. Reasons include:

- In most cases the designer did not have information on characteristics of the raw water over a reasonably long period of time, from which to assess ranges of the variables involved.
- The means and methods of determining design parameters from bench scale laboratory testing of raw water have not been well understood.
- Costs of obtaining good basic information may have been prohibitive.
- Characteristics of the raw water have changed in response to changing patterns of population, industry and agricultural activities in the drainage area.

Usually, the strategy in response to a lack of specific information has been to apply criteria that had proved satisfactory in other treatment plants. As a result, there may be many similar treatment plants originating from any one group of designers. It is important that treatment plant administrators and operators understand this background and seek to optimise the plant performance regardless of design, because significant improvements can be made.

The demand for water in all cities is increasing, such that most treatment plants are sooner or later pushed to their maximum and often severely overloaded. The result is poor water quality and high operation costs. Optimisation should seek the economical use of chemicals (which in many places are imported and very expensive) without detriment to water quality and it should improve operation to increase plant capacity or improve treated water quality, or both. Plant optimisation uses simple procedures and only a small amount of equipment, glassware, and laboratory space. Engineers and qualified, experienced operators can quite easily learn the procedures for initial assessment and for continued monitoring and refinement.

Bench scale laboratory testing is an effective method of optimising operation and uses settled water turbidity as a measure of clarification. This is valid because turbidity is related to the quantity of suspended solids in the sample. Other more precise methods of determining the microbiological and particle

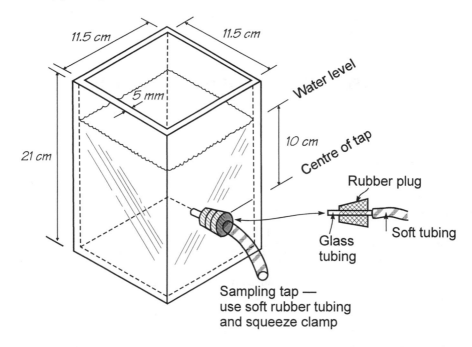

Figure 5.1   Two-litre jar for bench scale testing

content of a water sample are available but they are far more expensive, complex and time-consuming.

The bench scale work described here is rather like a miniature treatment plant operation. Judgement and care are necessary in scaling up the bench scale testing data, but experience shows that application of this type of information to the plant scale is very successful. The treatment plant, or a part of it, can be used to confirm test results without complicating excessively the routine operations. This economical and practical procedure has been carried out in many treatment plants.

The traditional jar test with simple modifications is the quickest and most economical way to obtain reliable data on variables that affect the treatment process and design parameters. The principle involved is clarification of 2 litre samples that have each been given specific treatments. The design and dimensions of the jar are shown in Figure 5.1. The jar is made of clear acrylic plastic of about 5 mm thickness so that it is sturdy, easy to transport, has good thermal qualities, and provides a clear view of floc formation. A tap allows samples to be taken with little disturbance. Depending on the treatment applied to each jar, clarification will proceed more or less rapidly, resulting in varying levels of turbidity.

Important parameters in designing or modifying a treatment process are surface loading, flocculation time and energy input. Intensity of mixing in loading of a settling basin is usually defined as the volume of water passing a unit of basin area in a unit of time. This is now most often measured as $m^3 \, m^{-2}$ per day, in preference to the more cumbersome gallons/ft^2/min.

Most conventional settling basins are designed for a surface loading of 30–40 m per day or about 2.1–2.8 cm per min. The square plastic jars are therefore designed to provide a distance of 10 cm between the water level and the centre of the sampling tap, which means that a sample drawn after 5 minutes of settling estimates the turbidity of a basin operating with a loading of 2 cm per min or 28.8 $m^3 \, m^{-2}$ per day. In other words, all the floc having a settling velocity of more than 2 cm per min will settle beyond the sampling tap after 5 minutes. It should be understood that a wide variety of particle sizes are produced in floc formation, and the larger particles settle more rapidly. Turbidity remaining in the effluent from a settling basin is mostly due to the smallest floc particles that settle very slowly. A reasonable settling time and plant capacity depends on their subsequent removal by filtration.

To determine accurately the settling characteristics of the treated sample, the first step is to measure the turbidity of the raw water. Then, after giving this a specific treatment (chemical dose, mixing and flocculation), the settling cycle begins and samples are removed after 1, 2, 3, 5 and 10 minutes of settling. These correspond to 10, 5, 3.3, 2 and 1 cm per min settling velocities. The first and last samples are the most and least turbid, but the amount of difference can vary widely according to the effectiveness of the applied treatment. There are many variables involved and so testing proceeds by varying only one condition between jars in each test. The best treatment is gradually identified by optimising each decision in turn. For example, if the coagulant dose is far from the optimum, then floc formation will be poor and clarification will be poor, even if other variables (mixing and flocculation) are close to optimum. The turbidity of the samples will reduce only a little during the test. By contrast, if the coagulant dose, mixing and flocculation are close to their best choices, the turbidity of early (1, 2 minutes) and later (5, 10 minutes) samples will differ substantially. Figures 5.2 and 5.3 show typical outcomes of initial and revised tests, from the example data in Table 5.1.

5.2 Design and process parameters

Important parameters which can be determined within a practical range of application are:
1. *Dispersion of coagulant, polymers and other reagents*
   - Effectiveness of the rapid mix based on limits of G applied by the stirring machine. Figure 5.4 shows the relationship of G to speed (revolutions per min (rpm)) for one instrument.

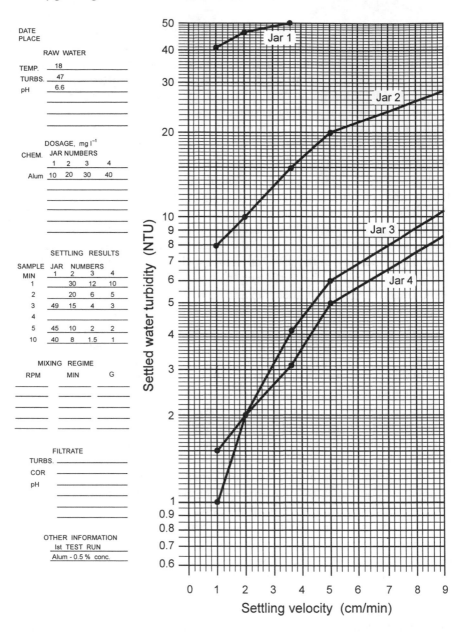

DATE
PLACE

RAW WATER

TEMP.   18
TURBS. 47
pH   6.6

DOSAGE, mg l⁻¹
CHEM. JAR NUMBERS
         1   2   3   4
Alum 10   20   30   40

SETTLING RESULTS

SAMPLE MIN	JAR NUMBERS			
	1	2	3	4
1		30	12	10
2		20	6	5
3	49	15	4	3
4				
5	45	10	2	2
10	40	8	1.5	1

MIXING REGIME
RPM   MIN   G

FILTRATE
TURBS.
COR
pH

OTHER INFORMATION
Ist TEST RUN
Alum - 0.5 % conc.

Figure 5.2 Jar testing to determine coagulant dose: first test run with a broad range of the test parameter. Test conditions and raw data can be recorded on the same sheet, as shown.

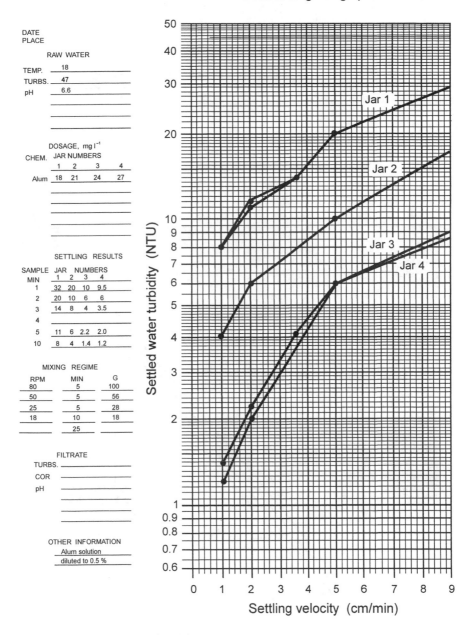

DATE
PLACE

RAW WATER

TEMP.   18
TURBS.   47
pH   6.6

DOSAGE, mg l⁻¹

CHEM.	JAR NUMBERS			
	1	2	3	4
Alum	18	21	24	27

SETTLING RESULTS

SAMPLE MIN	JAR NUMBERS			
	1	2	3	4
1	32	20	10	9.5
2	20	10	6	6
3	14	8	4	3.5
4				
5	11	6	2.2	2.0
10	8	4	1.4	1.2

MIXING REGIME

RPM	MIN	G
80	5	100
50	5	56
25	5	28
18	10	18
	25	

FILTRATE
TURBS.
COR
pH

OTHER INFORMATION
Alum solution
diluted to 0.5 %

Figure 5.3 Jar testing to determine coagulant dose: second test run with a revised, narrower range of the parameter under test

Table 5.1   Example of jar testing to determine the most effective coagulant dose (coagulant doses in units of mg l^{-1} Al$_2$(SO$_4$)$_3$)

Time (min)	Settling velocity (cm/min)	Turbidity (NTU) at coagulant doses (mg l^{-1}) of			
		10	20	30	40
First test					
1	10	56	30	12	10
2	5	54	20	6	5
3	3.3	50	15	4	3
5	2	45	10	2	2
10	1	40	8	1.5	1.0

Continued

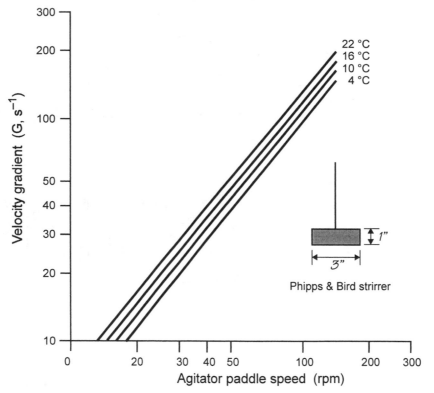

Figure 5.4 Relationship of applied velocity gradient to rotational speed and water temperature, for a stirrer which is commonly used in jar testing (Phipps & Bird) (Reproduced with permission from Water and Air Research, Inc.)

Table 5.1   Continued

Time (min)	Settling velocity (cm/min)	Turbidity (NTU) at coagulant doses (mg l⁻¹) of			
		18	21	24	27
Second test					
1	10	32	20	10	9.5
2	5	20	10	6	6.0
3	3.3	14	8	4	3.5
5	2	11	6	2.2	2.0
10	1	8	4	1.4	1.2

- Optimum dosages of the coagulants, polymers and alkalis as required.
- Effect of coagulant and polymer application — means and method, including dilution.
- Optimum timing and sequence of application of the relevant chemical reagents.
- Effect of pH on the treatment process.
- Effect of sludge recycling.

2. *Flocculation*
- Optimum total flocculation time.
- Optimum G value and time for each basin or optimum step-down values.
- Optimum compartmentalisation.
- Prediction of the effects of over-agitation and under-agitation on clarification.
- Tolerable velocities in transporting flocculated water between treatment plant units.

3. *Settling*
- Settled water turbidity and expected carry-over to filters at various basin loadings.
- Design criteria for settling basins, either conventional or high rate (tubes and plates), for nominal basin loading, entrance baffle and settled water removal system.

4. *Filtration*
- Floc load on the filter.
- Determining the possibilities of direct filtration.
- Approximate filtrate turbidity.

The important rule in running jar tests (and one of the reasons why they provide such good information) is that only one variable is assessed on each test run. The procedure then is to determine their best values one at a time, usually starting with the optimum coagulant dose.

5.3   Equipment and laboratory space

The following equipment is required for bench-scale laboratory testing:
- Stirring machine with variable speed control to about 150 rpm and with provision for at least four 2-litre beakers. Figure 5.4 gives G for 10–150 rpm with a 2.5–7.5 cm stirring paddle. A direct, positive, mechanical drive is recommended.
- Turbidimeter (light scatter type) of range about 0.01–1,000 NTU.
- pH meter.
- At least four 2-litre, square jars with siphons (Figure 5.1).
- Conversion curves to determine velocity gradient from stirring speed (Figure 5.4).
- Miscellaneous glassware or plasticware for holding samples, mixing reagents, making dilutions, etc. The lower risk of breaking high quality plasticware may be important.
- Pipettes of 1 ml, 5 ml, 10 ml, 25 ml and 100 ml capacity.
- Graduated beakers or flasks of 10 ml, 25 ml, 100 ml, 500 ml and 1,000 ml capacity.
- Stopwatch with a sweep second hand.
- Filter paper (Whatman No. 40).

The reagents needed are:
- 10 per cent $Al_2(SO_4)_3$ stock solution.
- 10 per cent $FeCl_3$ stock solution (for use each day, dilute stock solution to 0.1 per cent such that each 1 ml of coagulant solution will provide 1 mg of coagulant).
- Cationic, anionic and non-ionic polymers (manufacturers will provide samples for the testing). Make up polymer stock solution from dry reagent at 0.1 per cent concentration.
- Distilled water.
- Raw water.
- Lime solution, 1 per cent. Made up daily (may use another reagent for alkalinisation).

A small amount of bench space is adequate (about 3–4 metres of bench). This area should be provided with electrical outlets of appropriate voltage for the equipment and a sink with sufficient water for washing glassware and disposal of samples.

5.4   Testing in the laboratory for conventional treatment

5.4.1   Coagulant dose

The testing programme usually begins with determination of the optimum coagulant dose. If a treatment plant is already operating with the same raw water and coagulant under assessment, the first test run can often be made at about the plant dose. If new raw water or coagulant is under test, the range should initially be broad, narrowing in successive runs according to the results.

As with any laboratory work, all containers should be labelled explicitly and each measurement or action should be recorded on paper as it is taken, otherwise whole test runs may have to be repeated for the sake of small interruptions such as a telephone call. The procedure is as follows:

1. *Preparation.*
    - Check the turbidimeter for calibration over the anticipated range. Some of these instruments are calibrated separately for turbidity above and below 10 NTU.
    - Fill each jar to the 2-litre mark with raw water and measure the initial turbidity.
    - Carefully pipette the required dose of 0.1 per cent coagulant solution into a small beaker beside each jar.
    - If alkalinity is needed, measure out the necessary amount of lime solution for each jar and dose in advance, mixing slowly.
    - An example for the first test run might be 20, 40, 60, 80 ml of 0.1 per cent $Al_2(SO_4)_3$ set aside for jars 1 to 4 to provide 10, 20, 30, 40 mg l^{-1} of alum coagulant.

2. *Initial mix and flocculation.*
    - Pour the previously-measured coagulant dose from each beaker into its corresponding jar, with the mixing machine at maximum speed.
    - After 10 seconds of initial mixing, reduce speed to 100 rpm for 2 minutes, then reduce to 60 rpm for 3 minutes, then finally reduce to 20 rpm and continue for 15 minutes.
    - Just before the end of this 20 minute flocculation, drain each jar down to its 2-litre level to compensate for increased volume due to the 20 ml, 40 ml, 60 ml and 80 ml of coagulant.

3. *Settling and sampling.*
    - Two people are needed for sampling in order to avoid a serious time lag between sampling jars. With practice, each person can sample two jars at once, so that all four jars are sampled together.
    - Stop mixing to begin the settling cycle. It is good practice to discount about 30 seconds after mixing is stopped, while the rotary motion

subsides. This simulates time taken in the treatment plant for flocculated water to enter settling basins before settling begins.

- Start timing with the stop-watch set back to zero. Samples must be taken from each jar at 1, 2, 3, 5 and 10 minutes to provide settling velocity curves.
- Just prior to each sampling time, drain the siphon tubes (about 5 ml each).
- For the light-scatter turbidimeter a sample of 30–35 ml is sufficient, which can be taken quickly without lowering the water level in the jar too much.
- Take pH readings of each jar (those with higher dosages should have a lower pH).
- With four jars, there will be 20 samples for turbidity determination. Agitate samples carefully before pouring into the turbidimeter tube.

4. *Determining the best dose.*
- After the first trial run the process is repeated, using doses close to that which gave the best clarification (lowest turbidity).
- Turbidity of the 5 minute sample for a near-optimum dose should be about 2–5 NTU.
- When an acceptable coagulant dose has been found, as a compromise of effectiveness and economy, this should be used for all subsequent testing with the same raw water.

This testing procedure should be repeated for each coagulant that may be used in the water treatment plant. Table 5.1 shows examples of data obtained when searching for the most effective alum dose. In this example the most effective does was about 24 mg l^{-1}, which produced similarly effective clarification to the 27 mg l^{-1} dose but was about 10 per cent more economical in the use of coagulant.

5.4.2   Polymer dose

If the polymer dose is considered as an option to assist flocculation or settling it should be investigated next. The charges carried by the polymers determine their action. The positively charged cationic polymers act as coagulants and are used to supplement the coagulant dose. In some cases a small dose of cationic polymer during initial mixing can significantly reduce the required amount of primary coagulant. The feasibility of this procedure can be examined by bench scale testing in the laboratory. In addition to the quantity of polymer, the sequence of coagulant and polymer addition may be important, and tests can show which should be added first. Occasionally the polymer alone will form a satisfactory floc.

Non-ionic polymers are used to assist in the agglomeration of floc for settling. In this case the polymer must be applied after the floc has been formed, otherwise it will be less effective; but if applied too late there may not be enough time remaining for optimum floc formation. It is therefore important to test carefully when in the flocculation process to add such polymers. Experience suggests the best time is about 5 minutes after beginning the flocculation cycle.

The present authors have had little success with anionic polymers in coagulation and flocculation, whereas good results have been achieved with some cationic polymers in the coagulation phase, and excellent results with high molecular weight non-ionic polymers which assist during floc-building and settling. Polymers, however, are very expensive and should be used only when needed and effective. When evaluating polymers, the same general procedure and mixing regime is followed as for coagulants, using the coagulant dose that was previously selected. The stage at which polymer is applied will depend on the objectives and polymer type. They should be added as a very dilute solution (at most 0.01 per cent) in bench scale work, because low concentrations are required. Smaller volumes of more concentrated polymer solution are difficult to measure.

5.4.3 Examples of coagulant and polymer optimisation

Figures 5.5–5.12 show test results on waters from various parts of the world. The importance and relevance of information obtained from laboratory bench-scale testing is clear. With good judgement based on prior experience (perhaps by engaging a water treatment engineering consultant) more complex test regimes can be used to assess a range of options, as in Figure 5.5. Here a four-fold decrease in settled water turbidity from the current state (dip samples) can be predicted by an economical combination of lime, coagulant and polymer.

Less complex testing (the stepwise approach recommended here) is usually more efficient if the study is being done "in-house" by water department engineers and plant operators. Each issue can best be addressed in turn including coagulant dose (Table 5.1, Figures 5.2–5.3) and dilution (Figures 5.9–5.10); sequence of lime and coagulant addition (Figure 5.11); and polymer selection (Figure 5.6), dose (Figure 5.7) and timing of application (Figure 5.8).

Bench-scale testing which simulates plant conditions can also show whether clarification is reaching its full potential, given the prevailing raw water and chemical dosing. A need for attention to conditions in the treatment plant may be identified, as illustrated in Figure 5.12.

5.4.4 Flocculation conditions

Once dosages are determined, including dilution and sequence of application, the mixing time and energy input can be tested. This has such a wide range of possibilities that experience is very helpful in arriving at an optimum

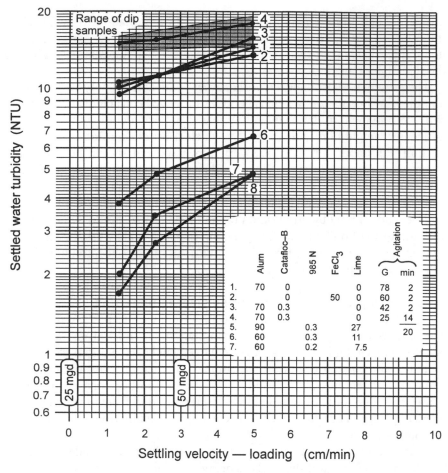

Figure 5.5 An advanced approach to jar testing in which lime, coagulant and polymer each vary. This may provide results quickly but requires considerable experience in jar testing, from which to judge the appropriate range of conditions. Jar 7 gives a good result and is economical (Reproduced with permission from Water and Air Research, Inc)

reasonably quickly. Thus, in deciding trial alternatives, a prior assessment of the plant flocculation system and improvement objectives is important. For new plants, the designer can specify a system that will produce the desired results, provided principles and experience are followed. If the objective is improvement of an existing plant, the system is there on site to analyse. Such plants will probably mix by vertical paddle, horizontal reel, oscillating paddle, or walking beam systems. Some of these are more readily adapted than others to provide a tapered energy input in the flocculation process.

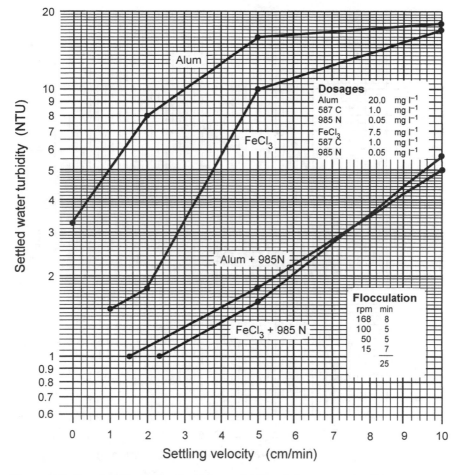

Figure 5.6 Results of jar testing for high energy flocculation, assessing the performance of a polymer (985 N). The polymer was very effective with both coagulants.

Every effort should be made to use existing equipment, but sometimes the old system must be abandoned. For example, if four compartments are already available, the bench scale work can be arranged to decide between mixing regimes which are compatible; 20 minutes flocculation would theoretically allow 5 minutes in each of four compartments, and hence the testing work could be orientated accordingly. In a horizontal reel system, the possibilities are broader because it can be made into a continuous channel with velocity gradient G values and time varying as water flows through the system.

Experience in sorting out this complex matter is fairly consistent for clarification of coloured and turbid waters. A short period of high energy agitation followed by a relatively long period of low agitation, either with or without a

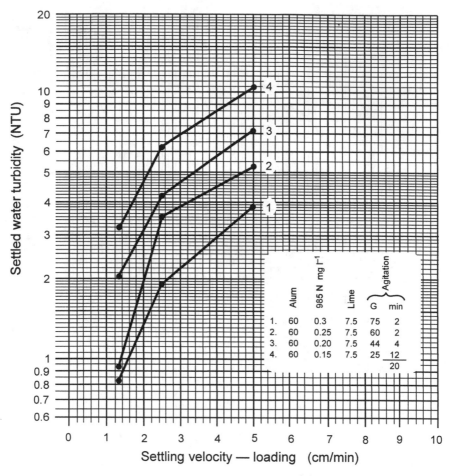

Figure 5.7 Results of jar testing to find the optimum dose of a polymer (985 N).
Clarification is directly related to the polymer dose (Reproduced with permission from
Water and Air Research, Inc.)

polymer, has been very effective. This is only a general observation and could
be less appropriate in specific cases, but provides a good place to start, otherwise inexperienced testers could easily get lost in the maze of possibilities.

The first test run should use the optimum coagulant dose, with addition of
0.1 mg l^{-1} of polymer, added just prior to the speed reduction at 5 minutes,
continuing for 15 minutes at 20 rpm to provide a total flocculation time of
20 minutes. The polymer should be chosen to assist floc bridging and
building (probably a non-ionic polymer of high molecular weight).

The objective of sampling is to construct a distribution curve of settling
velocity that will guide rehabilitation of existing flocculation and settling

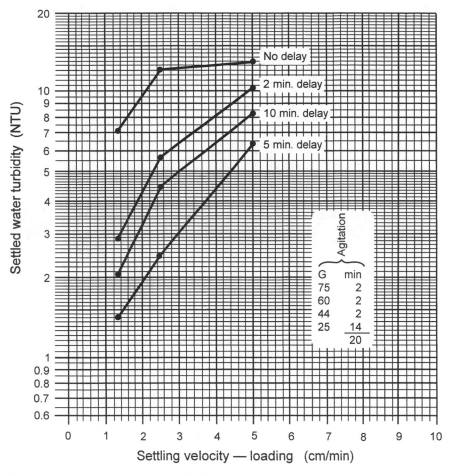

Figure 5.8 Results of jar testing to find the optimum timing of polymer application after start of flocculation. The polymer was most effective after the floc was well formed, and a further test run could be used to determine the optimum delay more closely (Reproduced with permission from Water and Air Research, Inc.)

basins, or the design of new ones. Settled water samples should be taken, as before, after 1, 2, 3, 5, and 10 minutes for turbidity measurement. Turbidity remaining in the sample is plotted on the vertical axis (on a log scale) against settling velocities in cm per min on the horizontal axis (linear scale), as illustrated in Figures 5.3 and 5.4. The flocculation cycle is then repeated with revised sets of conditions (the same approach as in coagulant testing) until the optimum mixing regime is determined. This will typically require more trials than were needed in the previous tests.

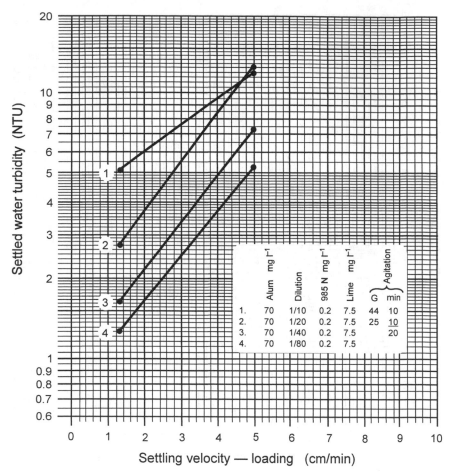

Figure 5.9 Results of jar testing to find the optimum dilution of coagulant. A constant amount of alum was diluted to a range of concentrations before application. Clarification is directly related to coagulant dilution (Reproduced with permission from Water and Air Research, Inc.)

The optimum mixing time remains to be determined by using the optimum coagulant and polymer dosages and optimum mixing regime, and stirring the jars for different times from 10 minutes to about 70 minutes (perhaps by 10-minute increments, using seven jars). Each jar is mixed for its specified time, then samples are taken at 1, 2, 3, 5, and 10 minutes for turbidity measurement. Each trial may need two runs if stirring cannot be provided for all of the jars. The data are recorded, but in this test flocculation time in minutes is plotted on the horizontal scale while turbidity is plotted on the

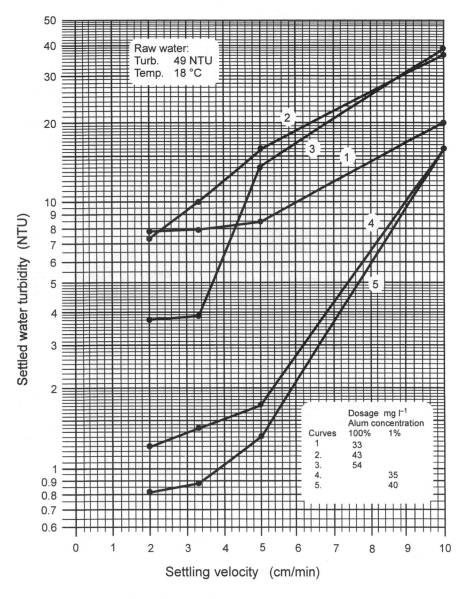

Figure 5.10 Comparison by jar testing with coagulant diluted and at day-tank concentration, using a surface water from east Asia. The dilute alum was consistently more effective than the concentrated alum.

vertical scale. Both are arithmetic scales. The clarification usually improves rapidly during the first 10–20 minutes, after which there is little change for a while, and eventually after 30–40 minutes the turbidity increases because of

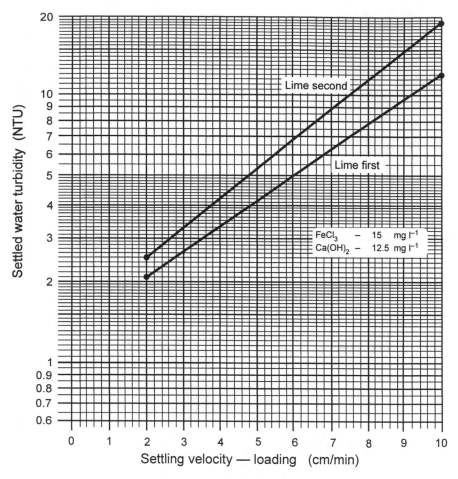

Figure 5.11 The effect of sequence of addition of lime and ferric chloride on clarification with water at 5 °C. The sequence of chemical dosing is often very important.

floc deterioration through over-mixing, which is more rapid with weaker floc (see Figures 5.12–5.15).

Figures 5.12 and 5.15 show the importance of flocculation time. Graphs of this variable are often plotted differently, with flocculation time on the horizontal axis and separate curves for each settling velocity (1, 2, 3, 5, 10 minute samples). An example is given in Figure 5.13.

5.4.5   Application of results

It is essential to note that the data apply to treatment variables under the conditions at the time of testing in the laboratory and treatment plant. For

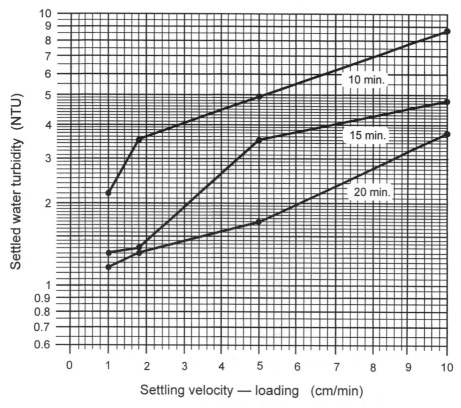

Figure 5.12 The effect of flocculation time on clarification, using alum and polymers. Flocculation time is very important for good floc formation (Reproduced with permission from Water and Air Research, Inc.)

successful application of bench-scale results to plant operation, conditions must correspond as closely as possible. Optimising flocculation time is a strong example, because its effect on clarification varies with temperature (Figure 5.14). Furthermore, if raw water characteristics such as turbidity, temperature or alkalinity change through the year, testing should be done periodically to revise estimates of optimum treatment.

The laboratory testing programme, including further aspects of plant operation, such as sludge return, results in a large body of valuable information upon which design or operation decisions can be made. Modifications may include physical structures as well as the process itself. Decisions can be made on treatment process design that would include:

- The most effective coagulants and optimum dosage range of each.
- The alkali dose range and periods of year when it is required.

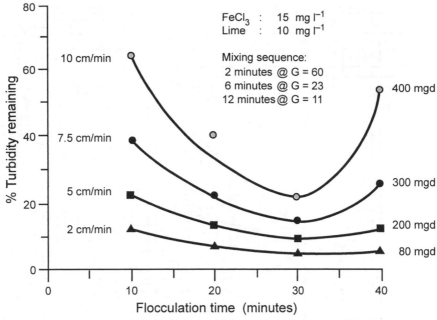

Figure 5.13 The effect of flocculation time on clarification at 5 °C, plotted by a method which is often used in examining this variable. Data for water from a large river. A less than optimum flocculation time produces poor results.

- The most effective polymer and doses.
- The timing and sequence of chemical application.
- The most effective process at extreme ranges of raw water variability.

Decisions on the physical process would include:

- The rapid mix, coagulant dispersion system.
- The flocculation system — hydraulic or mechanical (paddle, reel, turbine).
- The optimum flocculation time under extreme conditions.
- The optimum energy input under all raw water conditions.
- Flow control in flocculation — compartmentalisation or otherwise.
- Permissible flow velocities in ports, channel, pipes and the maximum velocity gradients to avoid break-up of the floc.
- Settling basin loadings under various conditions.

One further variable should be tested: that is the effect of pH adjustments in clarification. This is done by fixing all other treatment aspects and varying pH by lime addition, usually from pH 3.5 to about pH 9, and plotting titration curves of lime dose against pH. The curves in Figures 5.16–5.17 show how each water reacts differently and therefore how important it is for operators to know as much as possible about the characteristics of, and changes in, their water.

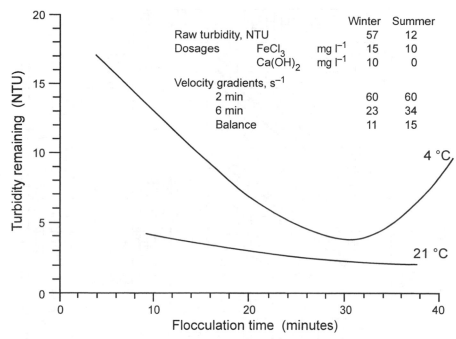

Figure 5.14 The effect of temperature on the relationship of settled-water turbidity to flocculation time, at settling velocity 5 cm per minute. A more than optimum flocculation time may produce poor results depending on the water and the temperature (Reproduced with permission from Water and Air Research, Inc.)

5.5   Testing in the laboratory for direct filtration

The potential economy, both in capital outlay and operating costs, makes direct filtration an attractive treatment process. The lower costs associated with reduced chemical consumption and sludge load have been emphasised less than initial capital savings in plant construction, yet represent continued savings through the life of the plant. Many raw waters are unsuitable for direct filtration. If the raw water is found to be suitable, preliminary bench-scale tests are needed, and pilot plant investigations should be made to establish plant design criteria.

The potential for direct filtration is decided by the amounts of coagulant and polymer that are needed to destabilise turbidity and colour, so that further reduction to required standards by filtration is economical. The maximum limits for turbidity and colour in treated waters vary among jurisdictions. No health-based guideline values for colour and turbidity have been proposed in the WHO *Guidelines for Drinking Water Quality* (Second Edition, 1993, published by WHO in Geneva), although it is stated that "*waters with colours*

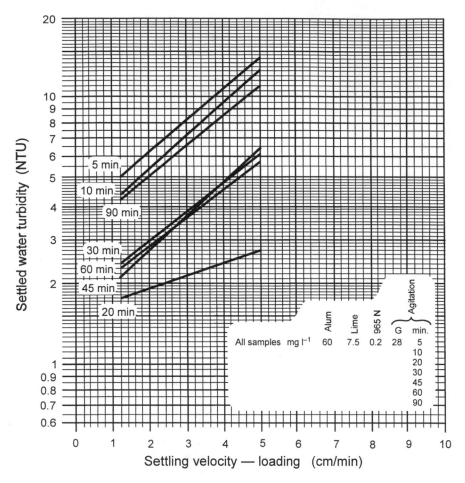

Figure 5.15 The effect of flocculation time on clarification. There is an optimum beyond which the quality of settled water deteriorates (Reproduced with permission from Water and Air Research, Inc.)

below 15 TCU ... [and] a turbidity of less than 5 NTU are usually accepted by consumers, but acceptability may vary according to local circumstances".

Low-dosage treatment depends upon destabilisation of the colloids in the water. Simple bench-scale testing can determine the destabilisation and filterability of colloids, identifying minimum treatment levels that produce low turbidity effluent through laboratory filter paper. Variables include the best coagulant, most effective polymer, optimum dosages, application sequence and mixing regime (i.e. a subset of those tested for optimising conventional treatment). Whatman No. 40 filter papers simulate removal by

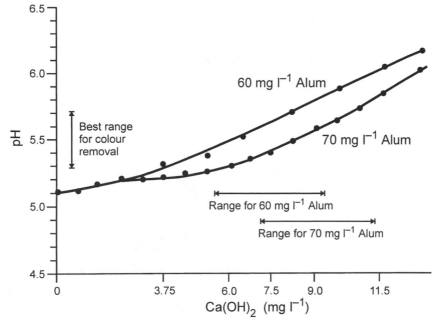

Figure 5.16 A titration curve of pH with addition of lime, used to find the lime dose which will provide best conditions for treatment with 60 or 70 mg l^{-1} alum. The curve itself is dependent on alum concentration. The pH range is important for good floc formation and economy.

pilot or plant filters closely enough. Filtration of each sample usually takes 2–3 minutes and papers are discarded after a single use.

Filter clogging is directly related to the floc volume that is loaded onto the filter, which in turn is related to coagulant dose. If the coagulant dose required for acceptably low turbidity and colour in filtered water is higher than 15–20 mg l^{-1}, economical filter performance is doubtful. When the required coagulant dose is less than 6–7 mg l^{-1}, with or without addition of a small dose of polymer, the water is an excellent candidate for direct filtration. The feasibility for waters needing an intermediate coagulant dose must be evaluated case by case. Raw waters of somewhat lesser quality may be accommodated by designing a filter with more storage and that will take higher loads.

As an example, the solids in water of turbidity 100 NTU (nephelometric turbidity units) occupy about 40 ppm by volume, if the particles causing turbidity have a specific gravity of 2.5. To coagulate and flocculate this turbidity requires at least 20 mg l^{-1} of a coagulant such as alum, producing about 5,000 ppm of floc and to avoid rapid filter clogging, such a load must

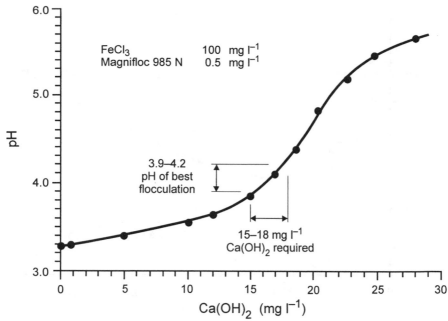

Figure 5.17 A titration curve of pH with addition of lime, used to find the lime dose which will provide best conditions for treatment with 100 mg l^{-1} ferric chloride

first be reduced by settling. If, however, this turbidity can be coagulated with 2–4 mg l^{-1} alum and/or a small polymer dose, the floc volume could be taken by a well-designed filter without unduly short filter runs.

Figures 5.18 and 5.19 give typical results of tests using various dosages of coagulant and polymer, and simply filtered through Whatman No. 40 paper. Some waters show a significant turbidity reduction by filtration alone (Figure 5.18), which is a strong indicator of suitability for direct filtration. In both of these examples a small polymer dose or somewhat higher coagulant dose was very effective, which also encourages further investigation of suitability.

Many bench scale tests can be carried out in a matter of hours. The optimum dosages can thus be determined economically in the laboratory, without first going to the expense of pilot plant design, construction and operation. If the results of bench scale testing are good, however, then pilot plant tests will be necessary to determine the design parameters of plant filters.

Typical pilot filter data from work on Lake Ontario water are given in Table 5.2. The duration of the filter run was inversely proportional to the alum dose, as discussed above in general terms.

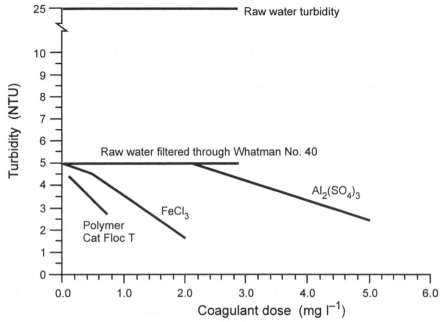

Figure 5.18 Results of a test for the feasibility of direct filtration. A low dose of cationic polymer or FeCl₃ was very effective. Filtration alone was sufficient to remove much of the raw water turbidity and this indicates that direct filtration would be suitable.

Table 5.2 Effect of alum dose on duration of the filter run at a Toronto pilot plant (Lake Ontario)

Turbidity of the raw water (NTU)	Alum dose (mg l⁻¹)	Average filter run (hours)	Turbidity of the filtered water (NTU)	Number of runs
1.8–3.5	5	35.9	0.16	12
2.9–3.3	20	9.8	0.14	14

5.6 Sampling for determination of treatment plant performance

After optimum values of the many variables have been determined in the laboratory through bench scale testing, the next step is to carry out sampling of plant performance for comparison.

5.6.1 Initial mixing of coagulant

Treatment plant testing can begin with initial dispersion of the coagulant in the raw water. First check that the dosing equipment is applying a correct dose. Once the coagulant volume, raw water flow and concentration of coagulant are known, the dose rate can be calculated. For example, if the solution

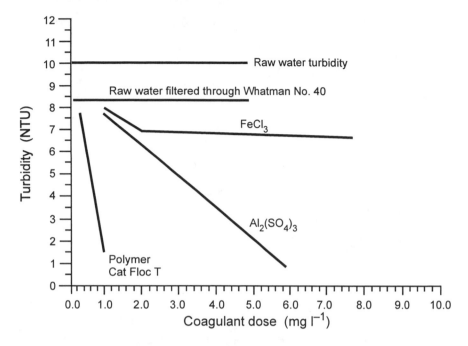

Figure 5.19 Results of a test for the feasibility of direct filtration, in south-east Asia. Filtration alone removed little of the raw water turbidity, but small coagulant or polymer doses were very effective. Some coagulants were more effective than others on any given water.

being applied is 10 per cent $Al_2(SO_4)_3$ or 100 mg ml^{-1}, the flow of raw water is 600 litres per min, and a sample of coagulant solution collected in one minute at the point of application is 300 ml, then:

Dose rate = 100 mg ml^{-1} × 300 ml per min / 600 litres per min = 50 mg l^{-1}

This measured dose should be within 5 per cent of the required dose, otherwise an adjustment is needed. The likelihood of a large dosing error increases with the concentration of the coagulant solution, because a small error in flow of the coagulant solution results in a large error in dose.

5.6.2   Flocculation system
After making sure that the coagulant dose is correct, the next step is to assess flocculation in the treatment plant, relative to the optimum results obtained in bench testing. Firstly, take samples of raw water between the initial mixing point and flocculation basins. Apply the optimum flocculation technique (time and energy input) to these in the laboratory as determined from bench testing and then take samples as in bench tests (at 1, 2, 3, 5, and 10 minutes

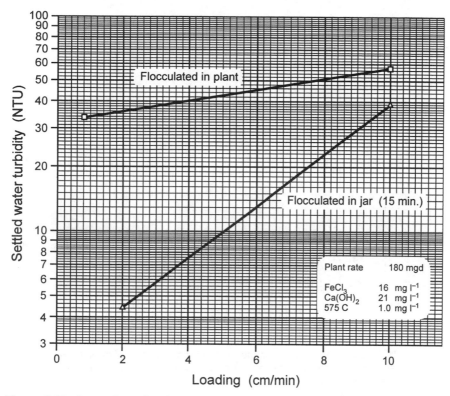

Figure 5.20 Comparison of settled water turbidity with flocculation in the plant and in the laboratory. The results indicate that the plant flocculation system needs improvement.

after settling starts) and measure their turbidity. Repeat several times to give settling velocity distribution curves. A reasonably good average and range can be determined: for example, the 5 minute turbidity reading might range from 4.5–6.0 NTU over 10 samples.

Second, take dip samples (using the square jars) from the outlet of the floc-culation basin, where the floc should be fully formed and ready for settling. This must be done with utmost care or the floc will break. The samples must not be agitated at all, but should be set down gently near the point of dipping. The bench scale settling procedure is carried out for several jars, taking samples for turbidity measurement at 1, 2, 3, 5, and 10 minutes. Settling velocity distribution curves can be constructed as before.

The curves for dipped samples, which reflect plant flocculation, are compared with those which were dosed and mixed in the treatment plant but flocculated in the laboratory. Usually, the better clarification is of labora-tory-flocculated samples, indicating scope for improvement in the flocculation system. Figure 5.20 shows the results of this procedure from an

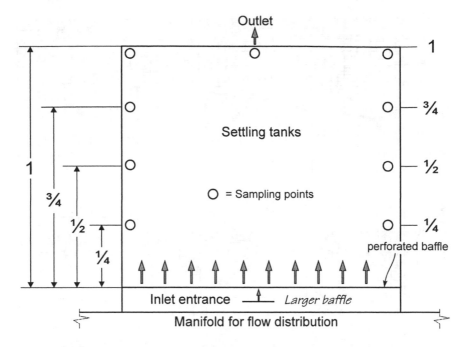

Figure 5.21 Location of sampling points to monitor distribution of turbidity within the settling basin. This will detect common patterns of failure, such as an increase of turbidity towards the outlet.

actual treatment plant, indicating the poor performance of the flocculator. Plans for improvement of flocculation were made and the modifications were carried out. Figure 5.40 (see later) summarises results at a plant in which doses, flocculation compartmentalisation and entrance to the settling basin were all modified. Improvements in the turbidity of settled water were spectacular, reducing from about 15 NTU to 1 NTU, which is a very good quality.

In most plants, poor flocculation is mitigated somewhat within the settling basin, which in such cases functions as a flocculation as well as a settling unit.

5.6.3   Settling basins

Samples for turbidity measurement should be taken from the settling basins at the same times each day for at least 10 days, perhaps at 10.00 h and 16.00 h; at 20–25 cm below the water surface from the outlet (centre and sides) and along each side at the quarter, half and three-quarter points from the basin entrance (Figure 5.21). Sampling from a given depth needs only simple equipment, such as a weighted bottle with its cork on a string that can be pulled when the bottle is in position. Sampling points can be marked on the basin wall for consistency.

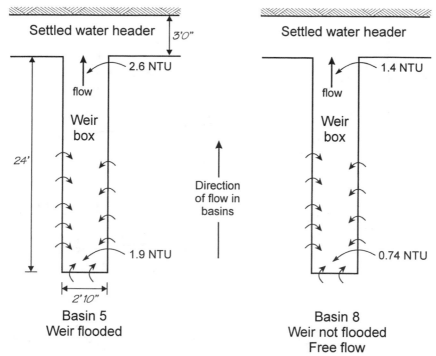

Figure 5.22 Turbidity variation around flooded and unflooded weirs at settling basin outlets. Flooded overflow weirs always contribute to poor basin performance.

Turbidity changes somewhat over the sampling period. Figures 5.22–5.27 present data from basins in several treatment plants with a common fault: the best settled water is near the entrance to the basin, or in its mid-section, while the poorest is at the outlet. This was due mainly to defective entrance design and short weirs (high overflow velocity) for settled water take-off systems. Where weirs are flooded this contributes further to carry-over of floc.

5.6.4   Filters

Samples of filtered water for turbidity measurement should be taken from each filter twice each day at the same hour and hourly when filter runs are less than 20 hours. For each filter, the number of hours since the last backwash should be recorded. This will indicate patterns of change during the filter run and identify any breakthroughs that might occur. It is important that each filter is sampled, because often just one or two filters are consistently the cause of poor filtered water quality. If pretreatment is satisfactory then the filters that are at fault can be identified for rehabilitation.

Filter rate and filter backwash rate can be readily determined by measuring the rise and fall of the water in the filter bed. This requires a wooden rod of

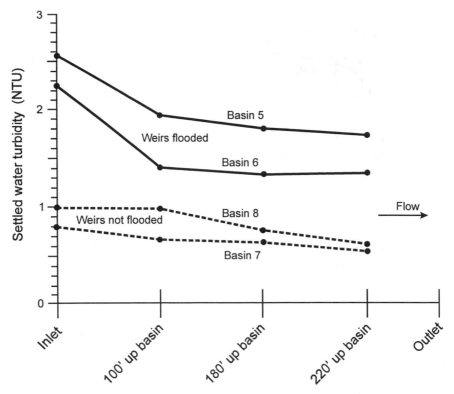

Figure 5.23 Variation in turbidity along settling basins with flooded and unflooded weirs. Flooded water outlets show poor performance.

3–4 m length, held vertically against the filter box wall and with marked intervals (e.g. Figure 5.28) or (better still) with a series of nails at 10 cm intervals having white-painted, upturned heads.

To measure filter rate, the drain and inlet valves are closed and the filtered water valve is opened. The time taken for the water level to fall 10 cm, 20 cm or 30 cm is then recorded. The area of the sand bed is known and, for example, if the bed is 5×10 m and the water dropped 30 cm in 2 minutes the filtration rate would be 7.5 m^3 per min (50 m$^2 \times 0.30$ m / 2 min) or about 216 m^3 m^{-2} per day.

For backwash rate, the rod is placed below the wash water overflow troughs, and the time for the backwash water to rise 10 cm, 20 cm or 30 cm is recorded. The calculations used are the same but the rates will be higher. For example, a 30 cm rise might require 0.5 minutes, and the backwash rate then would be 30 m^3 per min (50 m$^2 \times 0.30$ m / 0.5 min) or about 864 m^3 m^{-2} per day.

It is important to check how fast and how well the filters clean up during the backwash. This can be done by sampling backwash water at one minute intervals for 10 minutes after the first water goes over the wash water drain

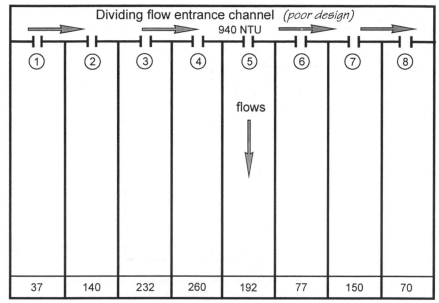

Figure 5.24 An example of poorly distributed water from a common channel, arising from poor design

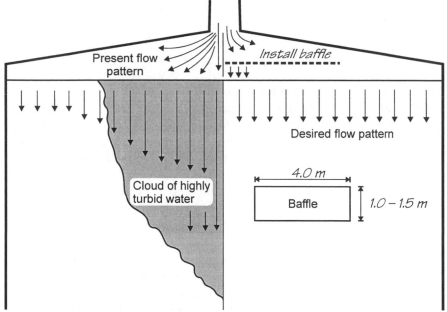

Figure 5.25 The effect of entrance design on settling basin performance. Installation of baffles improves performance

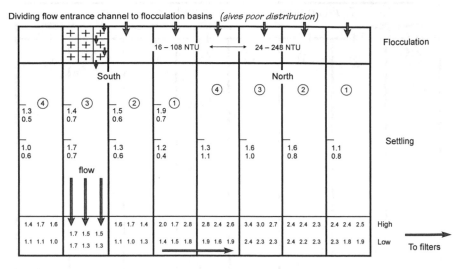

Figure 5.26 Turbidity measurements (five-day maxima, minima) through eight settling basins. Samples taken upstream of the basin outlet are always less turbid.

gullet and then measuring and plotting turbidity. Filters which are in good condition with an effective backwash should clean quickly (to 5–10 NTU in 4–6 minutes) and examples from filters with varied performance are shown in Figures 5.29–5.30.

5.7   Pilot plants

5.7.1   Fundamental considerations and design
Bench scale jar tests attempt to simulate treatment of water in the treatment plant and could therefore be considered pilot plants which operate on discrete batches of water. However, conventionally, pilot plants are understood as reduced-scale plant units that run continuously. Pilot plants have come to be regarded as complex, and expensive to build and operate. In some cases this is true but in most cases simple pilot plants can be designed, built and operated at a low cost.

The design of all treatment plants should be based on thorough bench scale and pilot plant information, including design of expansion units such as new filters, flocculation or settling basins, although this has been the case for only a small proportion of plants world-wide. Recently, pilot plant work has been used more often for design in industrialised countries. However, because the less developed countries have less money to spend, it is even more important that design of water treatment plants uses the best possible information from

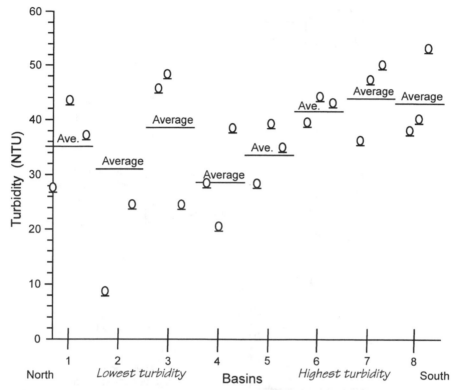

Figure 5.27 Elevation differences of launder inverts and averages for each basin (see also Figure 5.26). Differences in elevation of outlet weirs and launders can significantly affect basin performance.

bench scale and pilot plant testing. There are four main reasons why pre-design pilot plant work is omitted:

- Lack of understanding of the significance and need for pilot plant information.
- Urgency in getting new capacity into production to meet an unsatisfied demand.
- Unfamiliarity with the design, construction and operation of pilot plants.
- Failure to appreciate the economy that is gained by design based on sound, carefully derived information from pilot plants and full plant testing.

Reliable and applicable information can be obtained at very reasonable cost without special equipment or construction, e.g. from plant records and performance and from bench scale testing for the initial mixing of the coagulant, the flocculation system and the settling basins. Design data for filters, however, can be obtained only through operation of pilot filters. Continuous filter runs are required to provide information on head-loss, turbidity removal

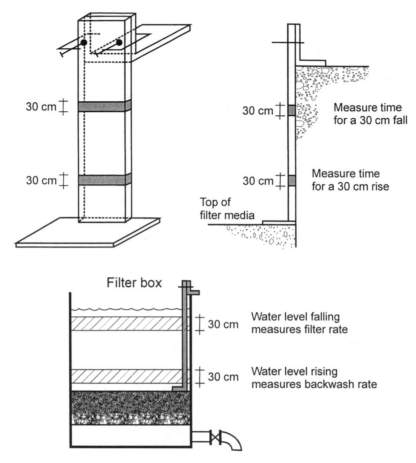

Figure 5.28 Simple gauge for determination of filter rate and backwash rate

and filter rates for specific media and pretreatment. Such filter runs may be as short as 8 hours or as long as 60–70 hours, and for this work a pilot filter with a pretreatment system must be constructed.

One of the best investments any water department can make is a pilot plant for the long-term study of water treatment processes. While these studies are usually carried out in relation to the impending design of plant expansion or new plants, operation over a long period of time is still needed to allow for changes in raw water quality. To obtain good information, the unit must be operated by qualified personnel and for whom it is their principal activity.

To build a pilot plant that will be easy to operate and will produce good data, some basic information and decisions are required. The first thing to decide is the size (flow) which is related to the number and size of pilot filters. These may be only 2.5 cm diameter or as large as several square metres, but

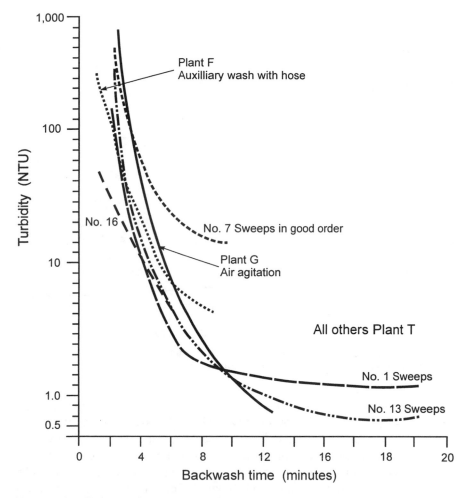

Figure 5.29 Filter cleaning rate during backwash at three treatment plants

are most practically about 15 cm in diameter. Very small filters bring hydraulic problems specific to small flows which are difficult to measure and control.

Although any pipe of 15 cm diameter can be used to construct a pilot filter, the expense of transparent plastic pipe is worthwhile. It is very important to observe floc penetration, plaque formation on the surface and in the media, media expansion during backwash, and the mixing of sand and coal in dual media filters. The part that needs to be transparent is that which contains the media and which provides a view of the surface with gravel and media expanded during backwashing. This is a total of not more than 1.5–2.0 m of pipe. The remainder can be made from any matching diameter pipe, but

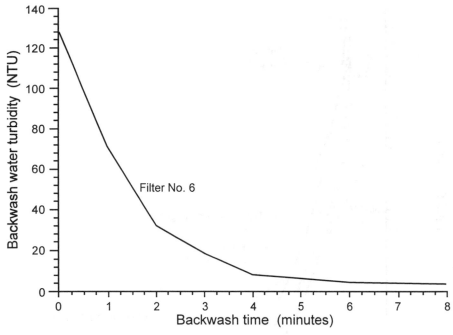

Figure 5.30 Filter cleaning rate during backwash, under auxiliary surface wash with a high-pressure hose. Filters should wash clean (i.e. < 10 NTU) in 5–6 minutes.

preferably made of PVC because of its light weight and the ease of jointing and handling. Figure 5.31 shows the typical form of such a pilot filter.

The preparation of raw water for pilot scale filtration must reflect conditions intended during operation. In many cases, the existing treatment plant can be used for pretreatment, using the best method determined in bench scale testing. Settled water can then be taken off directly for the pilot filter. If for some reason the existing plant cannot be used, or if a new plant is being designed, then small-scale mixing, flocculation and settling units must also be constructed. A hydraulic mixing method will usually be the best choice.

The decision of whether to go to the expense of a pilot plant can generally be made from results of the bench scale work, during which it is important to record information that will be directly relevant to the pilot plant. This includes optimum selection and dosages of coagulant and/or polymer, optimum flocculation parameters, and results of filtration using filter papers. For example, if doses required for destabilisation are very high and produce large amounts of floc, pilot testing of direct filtration is probably not worthwhile because the filters would clog too quickly. A pilot plant strategy based on bench scale testing may thus be very economical.

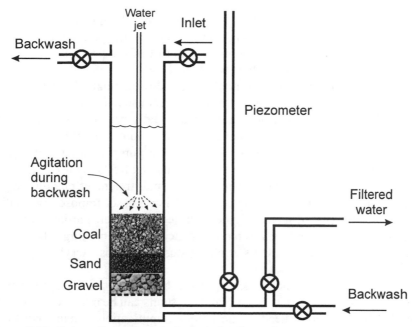

Figure 5.31 Schematic section of a typical pilot filter column

Before finally deciding on pilot filter testing, consideration must be given to filter design, which can strongly modify the acceptable floc load. Deep bed, coarse media filters (a metre or more in depth, of single or double media) can take high floc loads for quite long periods of time, which would quickly clog conventional filters. To use these deep filters, the designer must have some flexibility, which would be available in the design of new units. However, in rehabilitation of an existing plant, the concrete structure is in place thus limiting the designer's options.

The amount of water required in all aspects of pilot plant work is very small and therefore only small amounts of coagulant or other chemicals will be supplied. For example a 15 cm pilot filter operating at 300 m^3 m^{-2} per day requires 3.7 litres per min of raw water. Dosage of this flow at 15 mg l^{-1} then requires only 55 mg per min of coagulant, which is 11 ml per min of a 0.5 per cent (5 mg ml^{-1}) solution. This is a very small flow that is difficult to apply without special and accurate dosing equipment. The alternative is to treat a larger volume of water and drain the excess to waste, thus increasing the flow to make dosing much easier and more accurate.

If flow in the example above is increased from 3.7 litres per min to 37 litres per min, the required dose is then 110 ml per min of dilute coagulant solution. This is still a small flow but much easier to control than 11 ml per min. An error of 1 ml per min in the smaller dose may have a large effect on floc formation, but this error in 110 ml per min (less than 1 per cent) will not alter the result significantly. Discarding spare water during operation of the pilot plant is a small extra cost for improved control of the process. With more pilot filters operating, the wastage would be less.

There are two recommended methods for dosing small flows through a pilot plant: either an accurate dosing pump, or a constant head, gravity system with a Mariotte bottle. The dosing pump is usually preferred because of its accuracy and simple operation, but availability and cost may be significant problems in developing countries. A peristaltic pump with variable speed motor, pump head and plastic hose costs less than US\$ 500 in the USA, which is not substantial in the total cost for even a small treatment plant improvement. Some governments, however, restrict imports, or impose high duties that increase the cost of such equipment. Projects which are partially financed by an international bank or foreign assistance programme might be able to overcome this for some items such as dosing pumps, acrylic pipe and accurate valves.

If a pump system is not feasible, then a batch mix gravity system with Mariotte bottle can be used. Such gravity systems can be made to work well, provided the hydraulic head at the point of application remains constant. Pressure in the Mariotte bottle is always atmospheric and once the flow of coagulant is fixed, it will remain constant (see Figure 5.32).

Pilot plants provide valuable information on filter design and performance for whatever pretreatment is given. A typical well designed pilot plant will have three to six pilot filters which can be used to test three to six different filter options side-by-side. Results will be obtained more quickly with a larger set of filters, although careful, precise operation and control must be maintained throughout. A single pilot filter will provide excellent information if resources are scarce, but more time will be required to test a full variety of filter options. The options that may be tested using a pilot filter include:

- *Sand* — depth, effective size, uniformity.
- *Coal* — depth, effective size, uniformity.
- *Sand and coal combinations* — depths, effective sizes, uniformity.
- *Filter rates* — 150–900 m^3 m^{-2} per day (1.7–10.4 ml s^{-1} or 2.6–15.4 gallons/min/ft^2).
- *Filter control* — declining rate, constant rate.
- *Pretreatment* — conventional, direct filtration with coagulant mix, or direct filtration with coagulant mix and flocculation.
- *Backwash rates* — water wash alone (500–1,000 l m^{-2} per min with or without surface wash), water and air scour (air 50–60 m^3 m^{-2} per hour).

These options indicate the large number of variables (and their broad ranges) with which a pilot plant may deal. Even with several pilot filters operating, proper testing of these options requires a great deal of time. Water

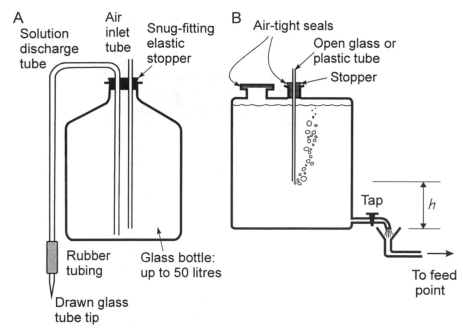

Figure 5.32 Mariotte bottle (left) and plastic box (right) for dosing coagulant and polymer

departments must allow for long-term operation of the pilot plant if the optimum treatment process is to be determined and monitored for change. This is especially true where the raw water quality changes significantly during the year.

The amount of money that any water department should invest in a pilot plant and its operation depends on the size and situation of the system. Cities everywhere are growing and as plants are getting older, new capacity will be needed sooner or later by expanding the output from existing treatment plants or by constructing new plants. Money spent obtaining sufficient information of high quality on the raw water and how it should be treated is well-invested in either case.

For small plants, one column filtering plant-treated water (settled, or only coagulated) may be satisfactory, whereas large cities may be justified in building larger pilot plants in anticipation of the future cost of new capacity. A new treatment plant in Los Angeles with a capacity of 15.5 m^3 s^{-1} operated its large pilot facilities for 3–4 years prior to design.

With one pilot filter, using pretreated water from an existing plant, the water department can experiment with filter media specifications to filter the water routinely produced in the treatment plant. This will provide information on filtered water quality, filter rates, length of filter run, final head loss,

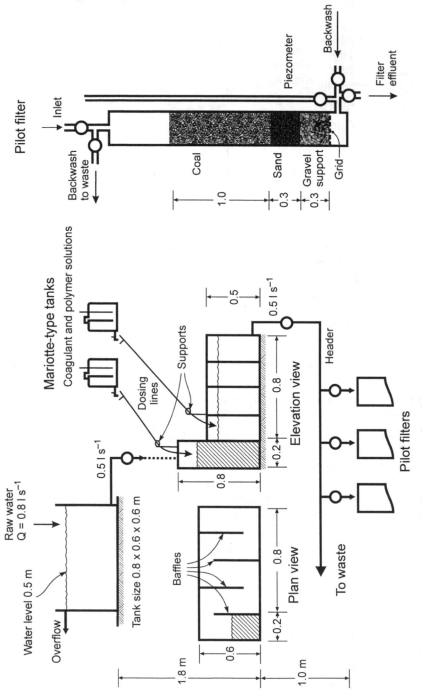

Figure 5.33 Layout of a simple gravity-operated pilot plant

required backwash rate and floc penetration under alternative options. This information can be used to select suitable specifications for future modification or construction. The layout of a simply built and operated gravity pilot filter system is shown in Figure 5.33.

Water can be siphoned from the settling basin outlet directly to the pilot filter column, which should be nearby to avoid long transmission lines that present problems of floc settling and changes in settled water quality. The flow in the siphon line should be at least 50 per cent greater than needed to supply the pilot filter, discarding the excess at the filter, which helps to keep the pipeline free from settling floc. Floc break-up due to increased stress in the pipeline is not important, because stress on the floc in the filter will be much higher than in the pipeline.

At the end of each filter run, media will have been compacted by the weight of the passing water. When the backwash starts, the tendency (even with filter columns of 15 cm (6 inches) in diameter) is for the entire media above the support gravel to rise as a piston. Some means must be provided for breaking up the media before it can rise more than a few centimetres. This can be done either mechanically or with high pressure jets of water inside the column. The agitation also helps break up the large pieces of plaque that tend to form in the filter media, especially when coal is used. Once the material is loosened, there is no longer a problem and the backwashing can proceed normally.

Information obtained from pilot plant work is most useful in relation to filter design, which unlike coagulation, initial mixing, flocculation and settling cannot be investigated reliably by bench scale jar testing. Each run provides data on head loss and turbidity removal at a specific flow rate with specific media. Figures 5.34 and 5.35 illustrate actual filter runs with raw water of about 60 NTU, successfully destabilised by a small dose (0.5 mg l^{-1}) of cationic polymer.

For the test shown in Figure 5.35, flow control was by declining rate, with the average flow rate fairly high at 285 m^3 m^{-2} per day. Head loss after 17 hours of the run was very reasonable at about 1.2 m. The pilot media were 26 cm of sand (effective size 0.67 mm, uniformity 1.07) and 85 cm of anthracite (effective size 1.40 mm, uniformity 1.57). Turbidity dropped rapidly below 1.0 NTU and stayed fairly constant at about 0.5–0.6 NTU during the filter run.

In another actual case at a conventional plant in western USA, the raw water turbidity was about 2.5 NTU and plant dosage was normally 2–2.5 mg l^{-1} of alum and 0.25 mg l^{-1} of cationic polymer. A study of plant records showed that very little was removed in the settling basins, and filters were responsible for about 95 per cent of the clarification. Four pilot filters were installed along the route taken by the water, receiving dosed water 20 seconds, 1 minute, 30 minutes and 2 hours after the rapid mix. Turbidity removal was almost the same for all pilot filters and in fact the first after

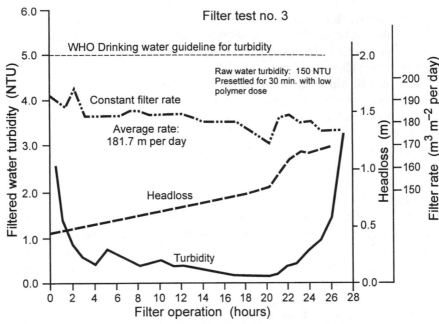

Figure 5.34 An example of pilot plant filtration with constant-rate flow control

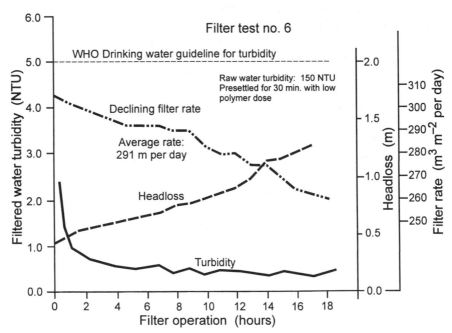

Figure 5.35 An example of pilot plant filtration with declining-rate flow control

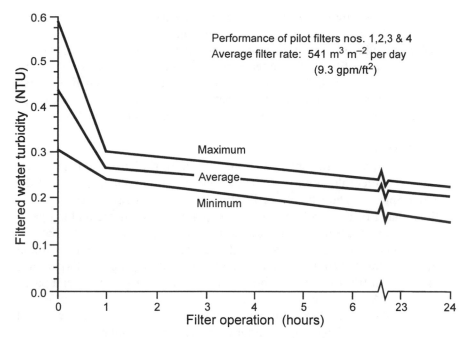

Figure 5.36 Long-term performance of pilot filters. Filters received water at several stages after mixing, and their consistency indicates suitability for direct filtration. With low raw water turbidity, filter rates can be five or more times the conventional rate (Reproduced with permission from Water and Air Research, Inc.)

mixing was consistently best by 0.02–0.04 NTU. The results of several months of daily filter runs are given in Figure 5.36. In this case, it was determined confidently that direct filtration would be a more appropriate treatment strategy.

Factors that should be considered in the design of a pilot plant include:
- How fast the information must be collected and assessed.
- Variation of raw water characteristics during the year.
- Provisions for pretreatment to be built into the pilot plant.
- Monitoring methods, whether continuous or grab sampling.
- Operation plans, whether continuous or interrupted daily for tests.
- The budget available for pilot scale testing.
- Types and quantities of monitoring equipment which are available and affordable.
- Data available from bench scale jar testing and the plant operating records.

In summary, the simplest pilot plant has one or two filter columns receiving water from the plant settling basins for filtration, with monitoring performed by grab samples and observations. The flow control may either be constant or declining rate. At the other end of the scale is a full, more expensive system for dosing and mixing of the coagulant with raw water, addition of polymer, reduced scale flocculation and settling, followed by a bank of four or more pilot filters. The latter plant will probably have dosing pumps, mechanical mixing and flocculation, a settling basin and each filter fitted with its own recording turbidimeter. In between these extremes, a wide range of alternatives is available to meet any needs for design information.

5.7.2   Operation

Even a pilot plant with only one column, filtering plant-produced water, requires considerable attention. The personnel requirement depends on the size of the plant, on how much prefiltration treatment is being performed, and on the number of filter columns. Once the plant is constructed and in operation, one trained technician may often operate it, but it will usually have to run 24 hours a day during a test. Although data can be gathered without continuous runs, they are required to simulate operation properly. Experience has shown that filtration data do not continue neatly before and after a pause in operation. At least two people, and perhaps more, are therefore required to measure and record filter information and monitor the various aspects of pretreatment through a complete filter cycle between backwashes.

The data collected during a pilot filter run usually include:

- Raw water turbidity (NTU) every hour through the period of the test run.
- Filtered water turbidity from each filter every hour during the test run.
- Total head loss every hour over the period of the test run of all filters.
- Chemical dose rates at the start of the test run and any changes during the test.
- Filtration rate during every hour of the test run.
- Specifications of media (depth, effective size, uniformity), size and depth of gravel.

5.8   Treatment plant as a pilot plant

Once satisfactory data have been collected from pilot plant operations, results can usually be tested further in a part of the treatment plant. Plant modifications which can be made to assess conclusions of bench scale jar and pilot scale testing are described below with two examples.

The first case was a large conventional treatment plant treating flashy, surface water for a large city in South America. The plant had originally been designed to provide $13.5 \text{ m}^3 \text{ s}^{-1}$ but more water was needed, so a programme of jar and pilot filter testing was undertaken. Settled water from the existing basins was routed to two pilot filters for testing sand and dual media options (Figures 5.37–5.39). The extent of short-circuiting in flocculation and settling systems was also examined.

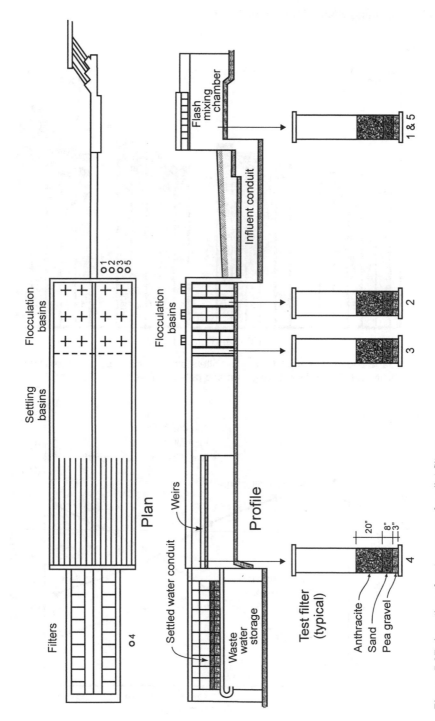

Figure 5.37 Location of water source for pilot filters

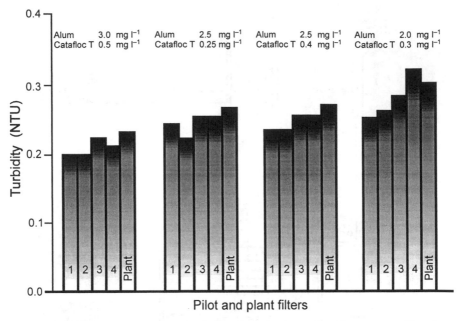

Figure 5.38 Performance of pilot and plant filters as determined by average filtered water turbidity with four different coagulant and polymer doses.

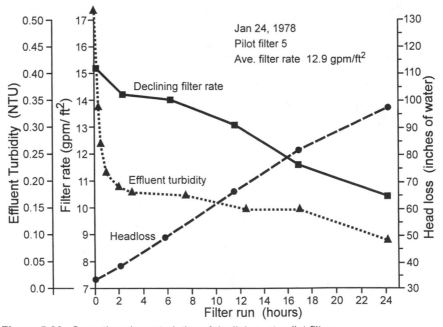

Figure 5.39 Operating characteristics of declining rate pilot filters

Investigations were done over an 18-month period. The suggested plant capacity could be doubled using the same treatment structure, subject to satisfactory performance of all stages in the treatment process. Flocculation, settling and transmission canals would be sufficient, and the sand filters were also suited for twice as much flow. The only serious hydraulic bottleneck within the plant would be the rate controllers on the existing filters. In addition, twice as much water would need to be supplied to the plant and distributed from it. In view of these optimistic conclusions, one flocculation system, one settling basin and one filter (a ninth of the plant) were modified to take twice their normal load. Filter alteration included removal of the venturi of the rate of flow controller to provide for increased flow rate. The filter sand was unmodified. This filter was operated at twice its normal load for about 6 months, monitoring head loss, length of filter runs, and filtered water turbidity.

The filtration rate was raised from 125 m^3 m^{-2} per day (1.44 mm s^{-1}) to about 250 m^3 m^{-2} per day. The turbidity of the filtered water was no greater as a consequence of this (0.3–0.5 NTU), although filter runs were reduced from 60–70 hours to 25–30 hours because of the greater loading with the same settled water (5–7 NTU). Given these results after a 6-month period of plant filter testing, there was no doubt that the existing sand filters could take the new higher loads.

One flocculation basin was modified to have six compartments instead of two. The two-compartment basin had a serious short-circuit under the hanging baffle, which meant that flocculation time was about 10 minutes instead of the designed 25 minutes. Short-circuiting was greatly reduced in the six-compartment basin, such that even with twice the flow about 12–13 minutes of flocculation was achieved, which allowed formation of good settleable floc. The next step was to improve the entrance to the settling basin. A temporary perforated entrance baffle was constructed to improve the flow pattern. However, during preliminary testing, the settling basins showed a tendency to form currents that carried water through the basin in 20 minutes instead of the 2 hours intended in the design. To improve settled water removal, launders of 20 m length were installed at the outlet end of the basin.

The preparation for this plant-scale testing took about two months. Although considerable expense was involved, this was quite justifiable with the possible economy of doubling plant capacity without new construction (bearing in mind the good outlook from smaller scale tests). Approximately US$ 30–40 million was the potential expenditure as opposed to the few thousand dollars actually expended.

Basin testing was carried out over two weeks (one day on, one day off). Turbidity of the settled water did not differ significantly between the settling basin with increased flow and the adjacent unmodified basin, further confirming the case for doubling plant capacity.

Design of the permanent plant modifications was initiated, based on results of progressive testing at bench scale, pilot scale and plant scale. Construction took 18 months and the treatment plant has been operating at a rate of about 29 m^3 s^{-1} for the past 25 years.

The second case was that of a colour removal plant in the southern USA. This was operating at its design capacity of 27 million gallons per day (1.2 m^3 s^{-1}) but about twice that amount of water was needed. Bench scale jar testing was initially carried out, along with sampling of flocculation and settling systems which indicated that there was considerable potential for improvement.

In jar testing, a number of options were tried and the best results were obtained by making several modifications to the usual operation of the plant:

- Changing from a cationic polymer to a non-ionic, high molecular weight polymer.
- Applying polymer in the flocculation basin, about 5 minutes into the flocculation cycle.
- Applying liquid alum as a 1 per cent solution, rather than at 50 per cent as supplied.
- Application of alum in a flow-through system, rather than a mixing basin with power-driven mixer.

The existing plant had eight propeller-type flocculation mixers, all mounted in one basin. It was therefore relatively easy to convert the basin into eight compartments, each with its own mixer. A new perforated baffle was installed at the entrance to the settling basin. For plant-scale testing, the set of modifications on one side of the plant were then as follows:

- The flocculation basin was further partitioned into an eight-compartment basin.
- A perforated entrance baffle was constructed in one settling basin.
- High molecular weight, non-ionic polymer was applied in the flocculation basin at about 5 minutes downstream from the beginning.
- Dilute alum was applied through a diffuser into the line from the raw water pump station.

The plant test was carried out over a period of 5 months with the modified side of the plant operating at twice the design rate. The results confirmed expectations, and the settled water turbidity of about 1–2 was lower than the turbidity before modifications. The plant itself had been used to confirm results of the jar tests. Figure 5.40 shows the results of bench scale and plant scale testing and the dramatic improvement in the modified plant.

The state agency controlling water plant operations approved the specified modifications for increased flow, and the plant has operated in this mode for many years. Approval for a higher load on the filters was not forthcoming,

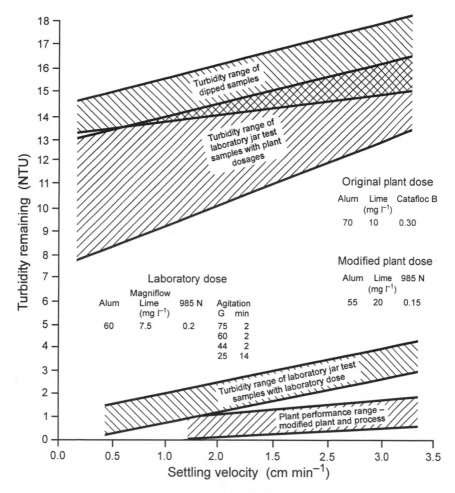

Figure 5.40 Performance comparison of existing plant conditions, plant dose and process, laboratory dose and process, and modified plant conditions. The significant improvements available from an optimisation procedure are clearly demonstrated.

although with lower applied turbidity they would have performed satisfactorily. In any case, the water department economised enough in the pretreatment to pay for the additional filters.

Chapter 6

DISINFECTION

6.1   Choice of process

There are many processes available for disinfection, some of which are physical or physico-chemical (e.g. thermal energy and ultra-violet, gamma, X or microwave radiation) and some are chemical (e.g. ozone, halogens and their compounds including chloramine, calcium and sodium hypochlorites). For economy, chemical disinfecting agents are most commonly used, especially chlorine and several chlorine compounds.

Some chorine derivatives can be carcinogenic. Those most relevant to water treatment are haloforms, which arise mainly by reaction of free chlorine with humic substances. Formation of haloforms is rather slow (from an hour to several days) depending on temperature, pH, the type and concentration of humic precursors, and the type of disinfectant residual.

Avoidance of haloform formation is relevant in choosing a disinfection method, especially if organic precursors are abundant in raw water. The first option would be to use alternative disinfectants but all processes studied so far have disadvantages. They either form organic by-products or a residual. Application is costly and control schemes are complex. Therefore removal of precursors before simple chlorination remains the most frequent choice.

Other considerations favour chlorination in the foreseeable future. Even when an oxidising agent is used in pre-treatment (for removal of algae, organic compounds, iron, manganese, taste and odours, or for viral inactivation) post-chlorination is able to provide a persistent germicidal residual which can be maintained throughout the distribution system. Chlorine is commercially produced even in less developed countries, costs of installation and operation are relatively low, and operators of water treatment plants have experience in using it.

6.2   Health risks of chlorine

There are a number of potential hazards that need to be respected during piping and assembling equipment for chlorine and during the handling of chlorine because it vaporises very rapidly at atmospheric temperature and pressure and is harmful even in small amounts.

Chlorine is a skin irritant, which may cause eye burns and damage to body tissues. Even at low concentrations the gas damages mucous membranes and the respiratory system causing throat constriction, coughing and pulmonary oedema. In severe cases of chlorine inhalation victims show excitement, restlessness, sneezing, salivation, retching and vomiting. Difficulty in respiration may cause death, and in less severe cases some permanent lesions may result. Therefore, safety is of great importance in the design and operation of chlorine facilities.

6.3   Design of chlorination facilities

Major points of concern in waterworks chlorination practice are effectiveness of operation and maintenance, ease of container storage and handling, and adequate control of the process. Although nearly every installation is different, there are some general guidelines that help in designing system components. These guidelines relate to containers; to feeding, metering and controlling equipment; to piping; to injectors; to housing; and to safety equipment.

The design needs to take account of chlorine consumption, which is a function of the flow to be treated at the required dosage. It is also necessary to consider whether to feed directly or in solution and to take account of the installation components which include the chlorine supply, chlorinator water supply, chlorine pipe lines, chlorinator and housing requirements. The system must be designed to incorporate strong precautions to avoid leaks and measures to control and contain leaks and to keep moisture out of all components that are required to stay dry.

6.3.1   Container, feeding, metering and controlling equipment

Chlorine is usually supplied as a dry liquefied fluid under pressure, in steel containers of 45.4 kg (100 lb), 68 kg (150 lb) or in 907.2 kg (2,000 lb, "ton containers"). The appropriate size depends on daily chlorine consumption and ease of supply. In general, the smallest cylinder is not used and the largest is chosen when consumption exceeds 250 kg per day.

Chlorinators measure only gaseous chlorine, which means liquid chlorine in the containers has to be vaporised before it reaches the feeders. Heat is required to change liquid chlorine to vapour, so some form of heating must be provided either from ambient air or in a specifically designed heat exchanger (a chlorine evaporator).

Proper operation of a chlorinator requires a minimum gas pressure at the point of gas withdrawal. The correct pressure depends on the type of feeder. For a direct-feed chlorinator this is about $1.41 \, \text{kgf cm}^{-2}$ (20 psi) plus the back pressure at the point of application; and for some vacuum-operated chlorinators (but not all) it is about $0.98 \, \text{kgf cm}^{-2}$ (14 psi).

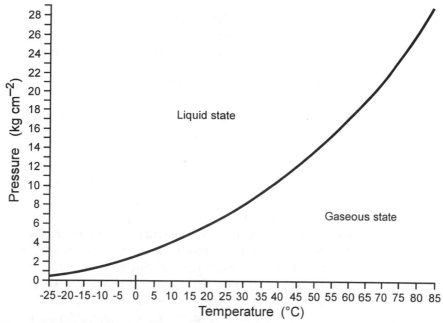

Figure 6.1 The relationship of chlorine vapour pressure to temperature

For rate of withdrawal (dependent upon transfer of ambient heat) several factors must be considered. The most important are ambient air temperature, air circulation rate, humidity of the air, the amount of liquid chlorine in the container, and the surface area of the container. Vapour withdrawal is more rapid with greater circulation of air, lower air humidity and lower liquid chlorine level in the container; but as the container is full at the start of the operation, and air circulation and humidity are variable, these three factors are generally disregarded. Calculations of sufficient accuracy need only account for ambient air temperature, container surface area, and minimum chlorine gas pressure at the point of withdrawal.

In the process of vaporisation of a liquid there is an equilibrium between vapour and liquid that depends on temperature and pressure. For chlorine the curve of equilibrium is shown in Figure 6.1 from which it is possible, given the minimum chlorine gas pressure necessary for operating the chlorinator, to determine the corresponding threshold temperature.

Taking into account the ambient temperature, the threshold temperature and the surface area of the container (called the withdrawal factor) the following equation is used to calculate the maximum withdrawal rate:

$$R = f(T_0 - T_1)$$

Where R = maximum sustained withdrawal rate (kg per day)
         T_0 = ambient air temperature (°C)
         T_1 = threshold temperature (°C) for the minimum required gas
         pressure
         f = withdrawal factor (0.64, 0.82, 6.63 for standard 45.4, 68 and
         907.2 kg (100, 150 and 2,000 lb) containers)
Given the required withdrawal rate, this calculation can indicate the number of containers to be connected to the same manifold without frosting under ambient air temperature.

The design must recognise pressure losses due to all components (valves, fittings, pressure gauges and piping) between the container and the chlorinator, because the minimum pressure required for the operation of the chlorinator must be provided at the chlorinator gas inlet.

There is a tendency to use the heat of the ambient air for chlorine vaporisation, principally in tropical countries or those that depend on importation of chlorination equipment. However, when the chlorine consumption requires more than four "ton containers" to meet daily demand, chlorine is withdrawn in the liquid state and a dedicated chlorine evaporator should be used. Whenever there is a sudden increase in the rate of flow, liquid chlorine in the evaporator vaporises rapidly causing a decrease of pressure in the whole system upstream of this unit. This in turn causes flashing gasification at points of highest friction loss, such as the entry and exit of valves and fittings, which hinders the flow of the liquid chlorine.

Longer pipelines and warmer liquid chlorine increase the time needed for the system to stabilise before restoration of normal flow. Consequently, the pipeline should be as short as possible. If the ambient temperature is high the pipeline should be isolated to shorten flashing and reduce the consequent pressure variation.

In the design of a chlorine storage area the following points should be observed:
- Containers should be protected from direct sunlight.
- Some means should be provided for the operator to know the chlorine consumption or when to switch over from containers in service to those in reserve.
- The distance from storage to the chlorinator should be as short as possible.
- Containers should be colder than the chlorinator, when operating in the gas phase.
- The container storage area should be above 10 °C because at lower temperatures the flow of chlorine is slow and irregular.
- Heat should never be applied directly to a chlorine container.
- Space should be allowed for at least two equal groups of containers (alternating service and reserve) plus unoccupied space for one container to help ease of operation.

- Outside storage areas should be fenced off for protection. Sub-surface storage should be avoided since chlorine, being heavier than air, will not rise to ground level in case of leaks.
- Cylinders should be handled with a two-wheel hand truck of the barrel pattern, and "ton containers" should be handled with an overhead conveyor for a manual or electric hoist and a lifting bar.
- Container hook-ups require an auxiliary valve (union or yoke) and a flexible connection.
- Note that temperatures of $-29\ ^\circ C$ or below may occur in chlorine systems, at which some metals become brittle. Chlorine is toxic, supports combustion of carbon steel and is very corrosive in the presence of moisture. Installations should use appropriate piping and components, be carefully executed, be well supported, be adequately sloped, anticipate pipe expansion due to temperature changes (use flexible connections) and should include a shut-off valve at the end of each line and expansion chambers in liquid chlorine lines.
- For chlorine gas or liquid service, at temperatures above $-29\ ^\circ C$ and pipe sizes of 19–44 mm (0.75–1.75 inch), pipes should be seamless carbon steel, schedule 80 (extra strong); the fittings and unions should be forged carbon steel, 907.2 kg (2,000 lb) cold working pressure; the valves should be forged steel globe type (363 kg, 800 lb), cast steel ball type (136 kg, 300 lb) or ductile iron plug type (136 kg, 300 lb). All joints should be screwed, welded or flanged. For screwed construction, the thread dope should be PTFE (polytetrafluoroethane) tape or paste, or cement prepared from litharge and glycerine, which will prevent joints from being disconnected. For welded construction, fittings should be of a slip-on type for pipes and weld neck for fittings. Gaskets should be of high-temperature compressed asbestos, ring type, 1.6 mm ($^1/_{16}$ inch) thick. Bolts and nuts should be heavy hex, carbon steel, with dimensions and threads as for specifications for flanges class 150. Valves should be cleared for chlorine service by the manufacturer, and their components should comply with the specifications recommended by the Chlorine Institute of the USA or other reliable organisation.
- Ventilation of the housing of containers should take into account the possibility of leaks, and provide exhaust fans capable of changing the air in less than 4 minutes. Care must be taken to avoid discharging chlorine where it can cause damage or injury.
- It is very important to clean all components before putting them into service because chlorine reacts with oil, grease and many other substances. Trichlorethylene or other chlorinated solvent can be used for cleaning the system. Hydrocarbons or alcohol should never be used because they may react with chlorine. Valves should be furnished already cleaned and should

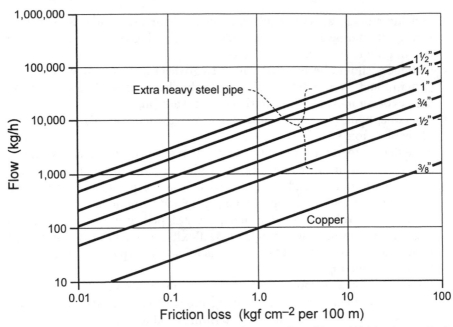

Figure 6.2   Friction losses for flow in steel and copper pipes (Darcy-Weisbach equation)

be tested with clean, oil-free dry air or nitrogen (dew point of no more than −40 °C) at 10.6 kgf cm^{-2} (150 psi) for seat tightness before installation. All piping systems should be tested at 21 kgf cm^{-2} (300 psi) before being cleaned with steam. While steaming, all condensate and foreign matter should be drained out, then all low sections should be drained; and while the system is still hot, air or nitrogen to above specifications should flow through the system for several hours to dry it. After drying, the system should be tested to 10.5 kgf cm^{-2} (150 psi) for leaks.

6.3.2   Piping

Flow characteristics of liquid and gaseous chlorine are different, but the Darcy-Weisbach equation can be used to determine pipe size for liquid chlorine. Figure 6.2 gives a nomograph for friction loss (pressure drop, as kgf cm^{-2} per 100 m of pipe) in several sizes of extra strong (schedule 80) steel pipe or 9.5 mm ($^3/_8$ inch) copper tube with respect to flow (as kg per hour).

As described above, sudden change in rate of flow causes liquid chlorine to vaporise. This phenomenon of "flashing vaporisation" causes abrupt pressure variation and forms extensions of chlorine gas which impede the flow of liquid chlorine. For this reason, the pipe diameter for a maximum pressure drop of 0.058 kgf cm^{-2} per 100 m is recommended for lines longer than 150 m and of 0.116 kgf cm^{-2} per 100 m for shorter lines.

Table 6.1   Average value of k (Darcy-Weisbach) for fittings and other appurtenances

Appurtenance	k value	Appurtenance	k value
Elbow standard screwed	2.00	Inlet	
Bend 90° flanged	0.45	Flush with wall	0.50
Bend 90° long radius	0.40	Inward projecting	1.00
Bend 45° screwed	0.40	Valve in open position	
Bend 45° long radius, flanged	0.20	Globe screwed	20.00
Tee-line flow screwed	0.90	Globe flanged	18.00
Tee-branch flow screwed	1.80	Gate screwed	0.35
Tee-line flow flanged	0.25	Gate flanged	0.30
Tee-branch flow flanged	1.00	Angle screwed	8.00
Sudden enlargement		Angle flanged	7.50
d/D $^1/_4$	0.95	Ball screwed	0.80
d/D $^1/_2$	0.60	Ball flanged	0.75
d/D $^3/_4$	0.20	Plug screwed	0.80
Sudden reduction		Plug flanged	0.75
d/D $^1/_4$	0.40	Diaphragm	2.30
d/D $^1/_2$	0.30		
d/D $^3/_4$	0.20		

Pipe length includes equivalent length calculated for fittings and unions, gauges, switches, valves, etc. An expression for the approximate determination of equivalent lengths, based on the Hazen-Williams equation, is:

$$L = 198 \times 10^{-6} C^{1.85} k D^{0.87} (P/\gamma)^{0.15}$$

Where L = equivalent length (m)
k = Darcy-Weisbach coefficient (see Table 6.1)
C = Hazen-Williams discharge coefficient (in this case taken to be 100)
D = pipe diameter (mm)
P = weight of liquid chlorine flow (kg per hour)
γ = saturated liquid chlorine specific weight (kgf m^{-3}) as shown in Figure 6.3

For gaseous chlorine flow, laws of fluid dynamics are applied. Friction losses in the system can be determined with sufficient accuracy using the appropriate formulae for flow in pipes. The Darcy-Weisbach formula (i) below can be transformed (ii) by various substitutions (iii):

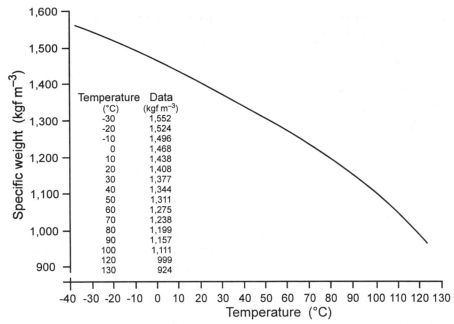

Figure 6.3 The relationship between specific weight and temperature for liquid chlorine

(i)
$$h_f = \left(f L v^2 \right) / \left(2gD \right)$$

(ii)
$$\frac{P_1 - P_2}{L} = \frac{0.0084\, f w^2}{\rho D^5} \qquad \text{or} \qquad \frac{P_1 - P_2}{L} = \frac{0.0084\, f w^2 V}{D^5}$$

(iii)
$$P = \gamma h_f \quad v = Q/s \quad Q = w/\gamma \quad S = \pi D^2/4 \quad \gamma = \rho g$$

Where f = coefficient of friction (diagram of Moody, as function of
Reynolds number)
L, D = pipe length, inside diameter (m)
v = flow velocity (m s^{-1})
g = 9.8066 m s^{-2}
P_1, P_2 = upstream, downstream pressure (kgf cm^{-2})
w = weight of flow (kgf s^{-1})
ρ = density or specific mass (kg m^{-3}, shown in Figures 6.4–6.5)
V = volume per unit mass (1/ρ)
γ = specific weight (kgf m^{-3})
Q = discharge (m^3 s^{-1})
S = pipe cross sectional area (m^2)
The coefficient of friction f as a function of Reynolds number and relative
roughness of the pipes (Σ/D) can be found using the diagram of Moody. The

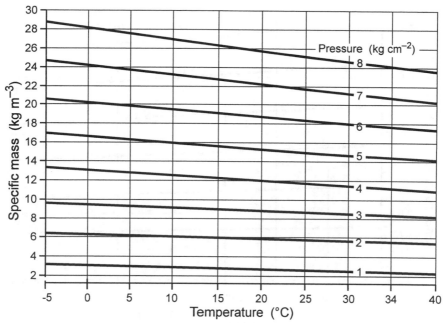

Figure 6.4 Density changes with temperature for gaseous chlorine

Reynolds number itself can be derived as shown below. For w in kgf s^{-1}, D in metres and μ in kgf s m^{-2}, $R = 0.1298\ w\ /\ (\mu\ D)$:

$$R = \frac{vD}{v} = \frac{\rho 4 w D}{\gamma \pi D^2 \mu} = \frac{4w}{\pi g D \mu}$$

Where v = kinetic viscosity μ/ρ

Dynamic viscosity μ with respect to temperature is shown in Figure 6.6. The following are roughness values Σ of the pipes employed in dry chlorine service expressed in millimetres: seamless carbon steel when new 0.02–0.04, after a year of service 0.06–0.10 and after several years (mild corrosion) 0.15–0.25; seamless copper or copper alloy 0.0015–0.010 (technically smooth); and PVC 0.06.

The minimum recommended nominal size for any dry chlorine piping system is 19 mm diameter ($^3/_4$ inch), and the best choice of material for chlorine under vacuum is PVC.

In treatment plants, four situations may usually be encountered which involve the passage of gaseous dry chlorine under pressure, flowing between:

1. Containers and the chlorine pressure reducing valve (CPRV).
2. The CPRV and the chlorinator.
3. The evaporator (when one is used) and the CPRV.
4. The CPRV and the chlorinator when an evaporator is used.

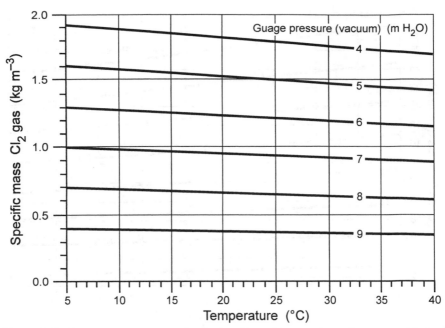

Figure 6.5 Gauge pressure according to density and temperature of gaseous chlorine

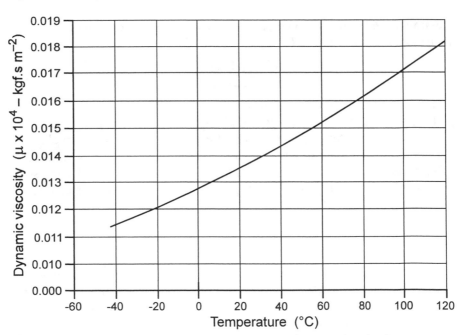

Figure 6.6 Relationship between dynamic viscosity and temperature for dry gaseous chlorine

In Case 1 the upstream pressure P_1 is vapour pressure at the container temperature, and P_2 is the CPRV outlet pressure. This valve improves chlorinator operation by reducing variable gas pressure to a desired fixed value and helps to avoid reliquefaction of the chlorine. It is normally set to an output of about 2.5 kgf cm^{-2}. Pressure available for sizing of the piping system is the difference between the output pressure of the CPRV and the vapour pressure at minimum ambient temperature. This difference is usually large and piping length is relatively short, so the calculated pipe size is small and the average gas flow velocity very high. There is a large pressure gradient (the pressure reduces quickly per unit of pipe length) and rapid expansion causes heat absorption from piping. To avoid ice formation outside the pipe and reliquefaction of chlorine, the velocity is usually kept below 10 m s^{-1}. The minimum recommended pipe size (diameter 19 mm) is satisfactory for flows up to about 3,000 kg per day.

In Case 2, the downstream pressure P_2 is the minimum operating pressure required by the chlorinator and P_1 is the output of the CPRV. Gas temperature drops with the abrupt pressure drop at the CPRV, so the density decreases and (for the same pipe size) the velocity increases. The new velocity has to be calculated from the minimum operating pressure required by the chlorinator at ambient temperature.

In Case 3, P_2 is output pressure of the CPRV and P_1 is gas pressure leaving the evaporator, which is c. 6.0 kg cm^{-2}. Gas temperature is about 37 °C so is 20.34 kg cm^{-3} for 7.0 kgf cm^{-2} absolute pressure and μ is 1.408×10^6 kgf s m^{-2}. After calculating pipe size by the Reynolds number and modified Darcy-Weisbach formula, the designer must verify the velocity.

For Case 4, P_2 is the minimum gas pressure for proper chlorinator operation, and P_1 is the output pressure of the CPRV. Usually, the line is short and is designed for average flow velocity not greater than 9 m s^{-1} using the ambient temperature for determining gas density. Pressure drop increases exponentially with increased flow velocity. A reduction of pressure increases the gas volume, which leads to further increased velocity and pressure drop (friction loss). For control of this problem, average flow velocity must not exceed the stated value.

6.3.3 Injector

For gaseous dry chlorine under vacuum (the flow of chlorine gas between the chlorinator and injector) care must be taken in design of the line when the injector is away from the chlorinators. Injectors are designed to provide 8.5 m H_2O vacuum but, due to leaks and other reasons, in practice at most 5.5–6.5 m H_2O vacuum is obtained. At least 3.5 m H_2O vacuum at the chlorinator is convenient for proper operation. Chlorinators actually operate at 1.7 m H_2O vacuum but twice this is preferable to draw out entrained air, CO_2

and molecular chlorine that may accumulate in the chlorine solution line. Total head loss in the line should therefore not be more than 2.1 m H_2O vacuum.

Several factors in the pipe-size calculations are dependent on the conditions of temperature and pressure. For example the friction coefficient as a function of Reynolds number depends on dynamic viscosity, which itself varies with temperature and the density varies with both temperature and pressure according to the PVT (pressure, volume, temperature) equation of state. Thus calculations are quite sensitive to the values of temperature and pressure, and for safety the most adverse conditions should be assumed in determination of pipe size. Analysis of the formulae for headloss shows that the appropriate extremes are the coldest ambient temperature, and the minimum vacuum provided by the injector, which might be 4.5 m H_2O vacuum.

The injector is not only responsible for the operation of the whole chlorine system, but also for dissolution of chlorine gas in the water which powers it. The concentration of saturated chlorine solution is about 3.5 g l^{-1}, depending on temperature and water characteristics. At higher concentrations the chlorine separates from the solution as molecular chlorine and either causes gas binding in lines under negative heads, or is released at the application point if this is open to the atmosphere. Therefore when an adjustable throat injector is used, a water flow meter should be installed so the operator can determine whether flow is sufficient for an unsaturated chlorine solution.

Four injector sizes are available, the smallest of which at 2.5 cm (1 inch) has a fixed throat, but interchangeable throats and tails can be selected according to the feed rate of chlorine. Injector efficiency dictates the flow and pressure of water for its satisfactory operation. Therefore, to decide appropriate water supply characteristics, the designer should obtain operation curves from the manufacturers of the injector.

The chlorine solution must be diffused as uniformly as possible into the body of water to be disinfected. This point is important in designing the diffuser. For pipelines flowing full, the diffuser should carry the chlorine solution to the axis of the pipe or to its diametrical plan, depending on whether pipe diameter is less or more than 0.60 m. In either case, the Reynolds Number of a pipe length for ten diameters downstream of the diffuser must be at least 2,500. For open channels the diffuser is either a series of nozzles in flexible hoses or a perforated pipe, inserted close to the bottom of the channel, designed so that holes are 0.10–0.20 m apart for flow at 3–4.5 m s^{-1} velocity. Figure 6.7 shows some characteristics of diffusers.

6.3.4   Housing

The design of a disinfection installation should take into account the arrangement of the chlorinator and its accessory equipment, heating and ventilation

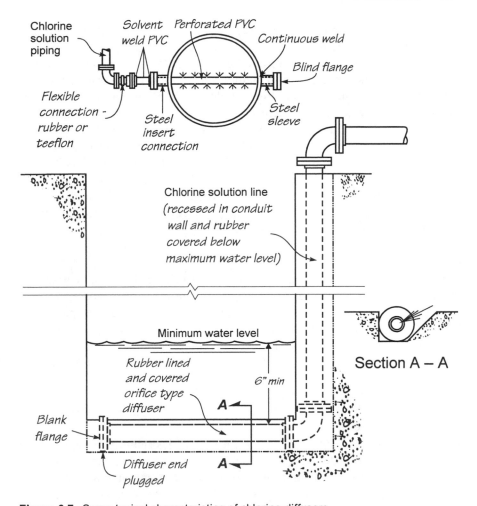

Figure 6.7 Some typical characteristics of chlorine diffusers

of the spaces for housing containers and chlorinators, and equipment for control and protection.

The chlorinator room should be kept at about 20–25 °C and must be heated if temperature falls below 10 °C. Container housing should be about 5 °C cooler than the chlorinator room itself, to prevent reliquefaction of chlorine on its way from containers to the chlorinators.

Natural ventilation may be adequate for chlorinators, provided the windows or ventilation arrangements assure cross-circulation of air and a complete air change every 15 minutes. In the case of severe leakage, air must be renewed in about 3 minutes by exhaust fans at or near floor level.

Vacuum-operated chlorinators have pressure-regulating valves upstream and downstream of the metering device, for accurate feeding. If system pressure varies due to a decreased or increased vacuum, a pressure–vacuum relief valve either vents or admits air to maintain the normal operating vacuum. In order to carry chlorine fumes to the outside in case of discharge, a PVC vent line must be provided for each chlorinator. These should not be longer than 7.5 m. The outside end should point downwards, covered by a copper wire screen to exclude insects and should have a slight downward slope (about 1 per cent) throughout, to drain any condensate. Each chlorinator has its own vent, so that operators can identify which unit is malfunctioning.

Chlorinators use a heater at the gas inlet to minimise deposition of possible contaminants, therefore an electrical supply must be available for this in the chlorinator room.

6.3.5 Safety equipment

Prevention of chlorine-related accidents is a priority, but it must be assumed that they will occur and the appropriate provisions must always be made. Gas masks and self-contained breathing equipment must be available to personnel in the vicinity of chlorine installations. Chlorine detectors and alarms must also be installed to give early warning of chlorine leaks.

Other points to consider in the design of chlorine installations are as follows:

- *Chlorine residual analyser.* This continuously measures free or combined chlorine in the water after contact. The analyser should keep a record, and requires an electrical supply.
- *Chlorine gas filter.* Chlorine has some impurities (mainly ferric chloride and chlorinated hydrocarbons) which may cause problems for piping and control equipment. To minimise maintenance of equipment, a gas filter should be used as close as possible to the last container and always upstream of any external reducing valve.
- *Evaporator.* Also called a vaporiser, this is used when it is impractical to vaporise chlorine by ambient temperature. It heats the liquid chlorine to its temperature of vaporisation at system pressure and superheats the gas slightly which helps to avoid reliquefaction
- *Chlorine leak detector.* This continuously samples the ambient air. Sensors rapidly detect chlorine gas at greater than 1 ppm by volume (3 mg m^{-3}), activating an alarm and light.
- *Container emergency kit.* This contains devices to counteract leaks that may occur in the container valve, the fusible plug, or the container itself.
- *Automatic gas switchover apparatus.* When a chlorine supply runs out, this changes to a new supply (although it is fitted with a manual shut-off) and it regulates container pressure to the best value for chlorinator operation. The

design of the apparatus is such that after switchover the container, which is nearly empty, continues to pass chlorine until it is fully exhausted. The apparatus also protects the system from impurities in the gas.

6.3.6   Examples

To complement practical recommendations made in this chapter, some examples of chlorine installations are shown in Figures 6.8–6.10.

Figure 6.8 shows an installation for the delivery of chlorine gas from "ton containers". The number of containers depends, as discussed, on chlorine withdrawal and ambient temperature. Low consumption in this case means that only one container is necessary, but a standby is installed to avoid interruption of chlorine flow during removal of the empty container.

The connection to the upper valve shows that the container is delivering gas to the system. An auxiliary cylinder valve (6) was installed to facilitate changeover of container (avoiding the entry of moisture into the flexible tubing and the release of some gas into the ambient air) and to shut off the chlorine gas flow in the case of the container valve being defective. The "ton container" is connected to the header valve (7) in the manifold by a flexible copper tube (13) at least 30 cm longer than the distance from the auxiliary cylinder valve (6) to the header valve.

Figure 6.8 also shows an angled steel bar (7.6 × 7.6 cm) for protection of the manifold; the filter (2) for removing inherent impurities in the chlorine gas; the pressure reducing valve (3) to prevent liquefaction of chlorine in the pipeline to the chlorinators; an alarm pressure switch (4) to warn the operator when it is necessary to switch from an empty container; and the pressure gauge (5) allowing the operator to monitor progress of the container more generally. The piping includes a branch to bypass the filter (2) and pressure reducing valve (3) when cleaning or maintenance is necessary. All equipment is isolated by means of line valves (1).

To move containers from the transport to the supply position (and to rotate them so that their outlet valves are in a vertical plane before connection to the supply system) a 2.5 ton capacity electric hoist (9) is required. The handling equipment also includes a lifting bar (8); a monorail (10) for the hoist trolley; wooden beams bolted to the floor (11) to keep containers dry; and end blocks (12) to hold containers in place.

Figure 6.9 shows a small three-chlorinator arrangement with a vacuum-operated feeder capable of applying 11–227 kg (25–500 lb) per day at two points (before and after filtration).

Chlorine gas from the supply system enters each chlorinator from a vertical down drop with a capped 30 cm trap leg (8) used as additional protection to trap impurities which remain after the filter. Each supply line has a valve (2). In case of malfunction, each chlorinator has its own vent line (6) to the outside, so

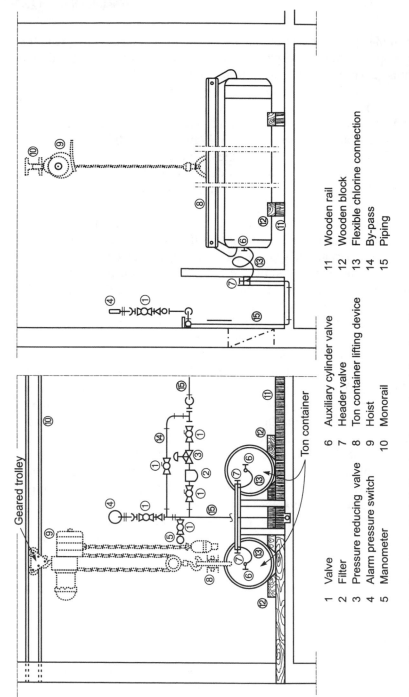

Figure 6.8 Example of a chlorine installation: housing and container handling

1	Valve	6	Auxiliary cylinder valve
2	Filter	7	Header valve
3	Pressure reducing valve	8	Ton container lifting device
4	Alarm pressure switch	9	Hoist
5	Manometer	10	Monorail

11	Wooden rail
12	Wooden block
13	Flexible chlorine connection
14	By-pass
15	Piping

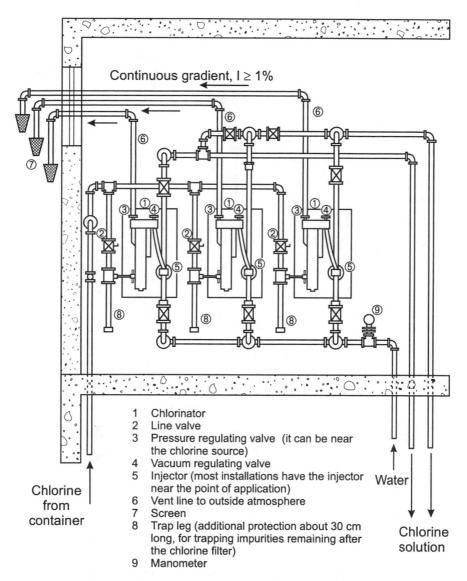

1 Chlorinator
2 Line valve
3 Pressure regulating valve (it can be near the chlorine source)
4 Vacuum regulating valve
5 Injector (most installations have the injector near the point of application)
6 Vent line to outside atmosphere
7 Screen
8 Trap leg (additional protection about 30 cm long, for trapping impurities remaining after the chlorine filter)
9 Manometer

Figure 6.9 Example of a chlorine installation: details of piping and metering

that operators can identify which unit is at fault. Water flow at the injectors (5) needed to produce about a 0.86 kg cm^{-2} (63.5 cm Hg) vacuum and to limit the chlorine solution to below 3,500 mg l^{-1} is determined indirectly by the manometer (9) in the discharge pipeline from the booster pump.

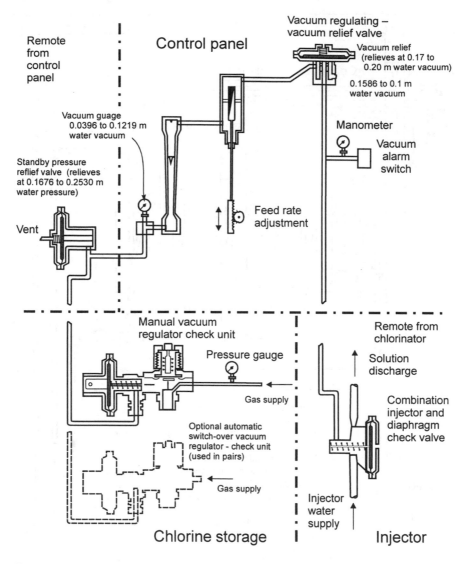

Figure 6.10 Example of a chlorine installation: overview of chlorine flow

The chlorinator in the middle is a standby unit for the two points of chlorine application: by proper valve operation its dosed chlorine solution can be sent to either chlorine diffuser.

Figure 6.10 is the flow diagram of a system with remote-vacuum arrangement and manual control. It shows the vacuum regulator check unit in the area of the chlorine containers and the injector near the point of application (that is, remote from the chlorinator).

Operation of the chlorination process begins by turning on the water supply to create a vacuum at the injector, which opens first the vacuum regulating valve and then the gas pressure regulating valve, allowing the chlorine to flow through the system from the containers to the injector. As soon as gas attains the fixed reference vacuum, the pressure relief valve closes, stopping the air flow into the chlorinator.

For a given temperature, the volume of a gas depends on its pressure, so precision of gas flow metering depends on keeping gas density and velocity as constant as possible at any rate setting. The chlorinators have a valve for this purpose, which automatically maintains the vacuum within precise limits, regardless of water flow in the injector, and the flow orifice is precisely grooved to assure quality of measurements either in manual or automatic control.

If the chlorine supply is interrupted, the vacuum relief valve opens and air immediately enters the chlorinator. An increase in vacuum within the chlorinator can be used to trigger an alarm, because lack of chlorine may cause serious air binding in the chlorine solution line.

An alarm switch is placed between the injector and differential-regulating valve. This switch is activated by either high or low vacuum conditions, to provide warning if the water supply fails, or if any hydraulic condition impairs the injector vacuum.

Chapter 7

OPTIMISING AND UPGRADING TREATMENT PLANTS

7.1　Applying new technology

Until recently there was no rational way of optimising the treatment process. Rarely, if ever, was the design questioned or any effort made to improve on what had always been done in plant operation. However, over the past 30 years, great progress has been made in understanding both physical and chemical features of the treatment process. New technology derived through objective research has opened the way for upgrading and optimising most existing plants, so that they may produce much more water of much better quality. This technology is based on:

- More complete understanding of the importance of the initial mixing of coagulant and raw water and of the means of accomplishing quick and total dispersion.
- The role of uniform distribution of velocity gradient throughout the entire water mass in both the initial mixing and the flocculation process.
- The importance of tapering energy input through the flocculation process for better floc formation and subsequent settling.
- A better knowledge of the application of manifold hydraulics in water distribution from a common channel or header to a number of parallel basins or pipes, and in collection of water in a manifold from a number of ports, orifices or other types of inlets.
- The design of entrance gates, dispersion of kinetic energy and flow distribution baffles of the settling basins and settled water removal system for most efficient clarification.
- The use of tubes and plates to accelerate the settling process.
- The use of polymers to improve settling and filtration and to reduce the coagulant dose.
- Hydraulic control of filtration for better quality, higher production, and protection from over-filtration because of operational errors.
- The use of single-media or dual-media deep-bed filters for improved quality and filter rates.
- The importance of thoroughly restoring filter operation by adequate backwashing.

Almost all plants in operation today have many defects in design and operation. Only a few recent plants designed with information from bench, pilot, and plant testing are likely to be less deficient. Optimisation and upgrading of water treatment plants (using new technology to recognise and correct defects) is very worthwhile for obtaining more and better water.

7.2   Optimising the treatment process and plant performance

The quickest and easiest way to obtain results is in the optimisation of the treatment process, which involves both physical and chemical aspects. The most common physical defects in water treatment systems are:

- *Related to dosing of raw water.* Inaccurate knowledge of raw water or coagulant flow, or of concentration of chemicals in batch tanks. Applying coagulant at high concentration, or at a point of low turbulence, or too far in advance of flocculation. Applying polymer in the wrong place. Failure to apply chemicals in their most effective sequence.
- *Related to flocculation and settling.* Over-flocculation and under-flocculation in terms of both time and energy input. Poorly designed inlets to the settling basins and to the settled water removal system.
- *Related to the filtration system.* Inadequate filter under drains and back-wash systems. Uneven backwash over the filter area. Backwash troughs not level. Flooding of the backwash troughs and gullet.
- *Related to disinfection.* Application of chlorine after upward pH correction. This means disinfection is likely to be poor, because chorine is not available in its most active form.

Common chemical defects are related to dosing of raw water: using coagulants or polymers that are ineffective, or are applied at the wrong dosages; or by dosing at the wrong pH.

Observations of plant performance and a study of plant records, if they exist, give a good initial idea of how the plant is performing for the raw water being treated. The next step is to decide on the testing programme. A minimum would be the bench scale testing described in Chapter 5, the results of which could indicate that pilot testing might also be appropriate. Testing provides information on the optimum treatment process and design parameters, which then must be applied to the existing plant in the most effective way.

As demonstrated in the examples of plant upgrading that follow, modification of plant units improves the effectiveness of each step in the treatment process, leading to a cumulative benefit as each step is passed along the treatment route. For example if raw water quality can be improved by changing the intake depth or location, the treatment process is immediately relieved, saving coagulant and improving water quality passing along the treatment route.

Good initial mixing with the correct amount of coagulant, followed by formation of a good settleable floc and an efficient settling unit, can significantly reduce turbidity of settled water, for example from 20 NTU to 3–4 NTU. Such an improvement has very significant results: the treated water quality will be much better; the concentration of solids, as well as bacteria and viruses, will be reduced in the same proportion; the filters will be relieved of an extra load of floc; and the main cause of filter problems will be eliminated. Once the filters are reconditioned they will produce excellent filtered water, and potential capacity of the plant may be greater.

In the chapters that follow, a description is given of three water treatment plants which were originally designed for 100, 1,000 and 20 l s^{-1} and have been upgraded to produce 250, 2,500 and 50 l s^{-1}. Depending on *per capita* consumption, but allowing a reasonable quantity especially for developing countries, these plants would serve cities or towns with populations of 100,000–125,000, of 1,000,000 and of 20,000–25,000 respectively.

Each of the examples takes a different bench scale result to apply to the design of the expansion. In all cases the most effective coagulant and polymers were selected based on jar tests and on direct filtration tests with filter paper. The full range of detailed calculations is not described for each treatment plant.

The most effective pH range was checked, although it is not usually necessary to adjust pH for effective treatment. Adding an extra chemical always increases the required work, cost and control, so every effort should be made to simplify rather than complicate the process. Unless there is a very pronounced need to increase alkalinity or to reduce pH, it is better to carry on the coagulation even though the pH may not be quite optimum. The floc formation and settling will not be significantly different nor will the required amount of coagulant.

Initial mixing of the coagulant and raw water is almost always defective in existing plants with coagulant generally applied where turbulence is insufficient for good dispersion. Such points are often on the surface of the raw water in a canal or in a mixing chamber. Moreover, coagulant is always applied at a high concentration rather than the optimum 0.5–1.0 per cent.

The next important parameters are the flocculation time, adequate uniform agitation, and energy input. The optimum details are readily found by bench scale jar testing and this has been used in the upgrading of each of the example plants. Optimum flocculation time is almost always within 18–30 minutes (depending on water temperature) and the best advice is to be conservative, which means that 25–30 minutes is usually selected as flocculation time.

The mixing energy is always best applied in a tapered manner. Usually the first few minutes of flocculation can be done with a high velocity gradient G of perhaps 80–100 s^{-1}, from then to about 12–15 minutes the gradient is reduced to 40–50 s^{-1}, then reduced again to 25–30 s^{-1}, and for the final 10–15 minutes a gradient of 18–20 s^{-1} is arranged. The flocculation basins must fit these parameters. In basins with hydraulic mixing this tapering is easy to accomplish by design. However, where there is mechanical mixing in separate basins details of tapered input must be adapted to the number of basins. Both hydraulic and mechanical flocculation are covered in the upgrading examples.

The velocity gradient in the last part of the flocculation system is accounted for in design of the perforated baffle wall at the entrance to the settling basin. The gates and ports in the entrance wall are designed to avoid breaking the floc that has been formed in an environment of around 20 s^{-1} velocity gradient. Sometimes, to attain the head loss required, a slightly higher gradient must be used in the ports. Because water flows through the ports in a second or less, there is little danger of floc damage and a velocity gradient of 30 s^{-1} in the ports is tolerable.

The settling basins are loaded to conform to the settling velocities which gave the best results in jar testing. A series of tests will give a range of settling velocities that produce settled water of 3–4 NTU. With the correct coagulant dose, and good initial mixing and flocculation, the optimum settling velocity will often be in the range 3.0–3.5 cm per min, and depending on the raw water perhaps 2.5–4.0 cm per min. In any case, existing basins or new ones are loaded subject to a margin of safety. If testing gave a range of 2.8–3.3 cm per min for satisfactory turbidity removal, the best practice would be to use a velocity of not more than 2.8 cm per min.

Another role of bench tests is to examine the possibility of direct filtration. If turbidity of coagulated water is reduced significantly with filter paper only then direct filtration may be feasible. Then if raw water turbidity and colour are favourable for enough of the year, the conventional plant can operate in direct filtration mode during such periods. This possibility is included in one of the examples of upgrading given.

7.3   Design information
Designs for upgrading (incorporated into the three examples of plant optimisation) use, for the most part, a common set of information and expressions for hydraulic and other calculations.

7.3.1   Weirs
Rectangular, sharp crested, aerated, with no end contraction:

$$Q = 1.838 \, l \, h^{3/2}$$

Where Q = flow (m^3 s^{-1})
           l = length (m)
           h = depth over weir (m)

V–Notch or triangular:

$$Q = 13265 \tan(\theta/2) h^{2.47}$$

Where Q = flow (m³ s⁻¹)
        θ = notch angle or angle of triangular weir
        h = head (m)

7.3.2 Channels

$$Q = Av = \frac{AR^{2/3} I^{1/2}}{n}$$

Where Q = discharge (m³ s⁻¹)
        R = hydraulic radius (m)
        I = slope (m m⁻¹)
        A = area (m²)
        n = roughness factor (concrete 0.013)
        $v = \left(R^{2/3} I^{1/2} \right) / n$ velocity (m s⁻¹)

7.3.3 Ports

$$Q = C_d A \sqrt{2gh}$$

$$C_d = 1 / \sqrt{15 + Kl/d}$$

Where Q = discharge (m³ s⁻¹)
        C_d = flow coefficient
        A = area (m²)
        K = roughness coefficient (Darcy's)
        d = diameter of port (m)
        l = length of port (m)
        h = head on the port (m)

G values for ports can be obtained from:

$$G = \frac{d_h C_d}{s} \sqrt{\frac{\pi}{8v}} \sqrt{\frac{V^3}{l_1}}$$

$$= (d/s) \sqrt{\pi/8v C_d^2 X} \left(V^{3/2} / X^{1/2} \right)$$

$$= K \left((d/s) V^{3/2} / X^{1/2} \right)$$

Where d = diameter of port or hydraulic diameter (m)
        s = distance on centres of ports (m)
        v = kinematic viscosity (m² s⁻¹ ...$v_{20\ °C}$ = 1.007 × 10⁻⁶)
        X, l_1 = length of jet path before merging (m)

V = jet average velocity in the ports (m s^{-1})
C_d = flow coefficient, obtained either in hydraulic tables or in the expression below:

$$C_d(\text{submerged}) = 1 \big/ l \sqrt{\Sigma\mu + \Sigma K\left(l/4R_H\right)}$$

Where $\Sigma\mu$ = local losses (coefficient of entrance + outlet losses)
R_H = hydraulic radius (DH/4)
K = friction factor (from the diagram of Moody)

7.3.4   Distributing manifolds
Flow in channel (Manning formula):

$$G = \sqrt{\gamma/\mu}\; I^{1/2} V^{1/2} = \sqrt{\frac{\gamma}{\mu n}}\; R_h^{3/4}\; I^{3/4}$$

Where R_h = hydraulic radius (m)
V = velocity (m s^{-1})
n = roughness coefficient ($n = 0.013$ for concrete)
γ = specific weight (998.23 kgf m^{-3} at 20 °C)
I = slope (m m^{-1})
μ = dynamic viscosity ($\mu_{20\,°C} = 1.03 \times 10^{-4}$ kgfs m^{-3})

7.3.5   Hydraulic sludge removal
Minimum pipe size is 200 mm (8 inches) and emptying time of a basin or receptacle is:

$$t = 2V \big/ \left(C_d A \sqrt{2gh}\right)$$

Where t = emptying time (s)
V = volume (m^3)
C_d = coefficient (0.65 in this case)
A = area of drain (m^2)
h = water depth (m)
g = acceleration of gravity (9.81 m s^{-2})

7.3.6   High rate settling in tubes or plates
Yao formula is:

$$\left(V_s/V_o\right)\left(\sin\theta + L\cos\theta\right) = S_c$$

Where V_s = settling velocity of particle to be removed (same units as V_o)
V_o = average flow velocity in tubes or plates (same units as V_s)
θ = angle of inclination of tubes or plates from the horizontal
L = relative length of the settling system (length/width)
S_c = parameter for each type of settler. For example circular

tubes ($S_c = {}^4/_3$), rectangular tubes ($S_c = {}^{11}/_8$), parallel plates ($S_c = 1$) or shallow open tray ($S_c = 1$)

7.3.7   Collecting systems

Troughs or channels with uniformly varied flow (horizontal rectangular):

$$\text{Water profile} = \left(x/l\right)^2 = \left[1+1/2F_o^2\right]\left[y/y_o\right] - \left[\left(1/2F_o^2\right)\left(y/y_o\right)^3\right]$$

$$\text{where} \quad F_o^2 = \left(q^2 l^2\right)/\left(gb^2 y_o^3\right) = Q^2/\left(gb^2 y_o^3\right)$$

Where F_o = Froude number (for free flow outlet $F_o = 1$, or else calculated as above)
l, b = length, width of channel (m)
g = acceleration of gravity (9.81 m s^{-2})
y, y_o = water height at x metres from the origin or upstream end, and at the outlet
q = discharge per unit of length (m^3 s^{-1} m^{-1})

7.3.8   Power for flocculation

For each value of the velocity gradient G (s^{-1}), the power applied to the water can be obtained from:

$$P = \mu V G^2$$

Where P = power applied to the water (kg m s^{-1})
V = volume of basin (m^3)
μ = dynamic viscosity ($\mu_{20\,°C} = 1.03 \times 10^{-4}$ kgfs m^{-3})

Applying power with a propeller:

$$\text{Power Number} = N_p = Pgc/N^3 D^5 \rho$$

N_p is usually represented by K but is obtainable from a graph as a function of the Reynolds Number, which for propellers is:

$$R_e = \frac{D^2 N_p}{\mu}$$

Where gc = Newton's law conversion factor ($\pm$ 9.81 kg m/kgf $\times$ s^2)
N = number of rotations per second (rps)
D = diameter of propeller (m)
ρ = density (specific mass, kgf $\times$ s^2/ m^4)

The correspondence between the Power Number and the Reynolds Number varies according to the type of propeller and is determined experimentally.

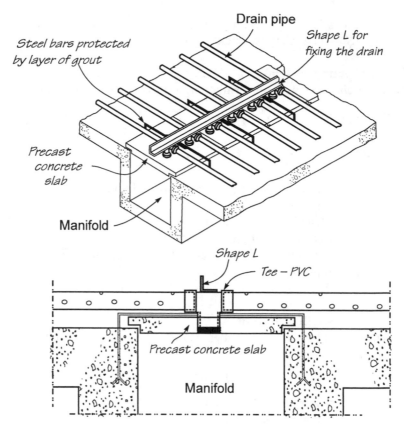

Figure 7.1 Layout of header manifold and laterals in a filter bottom

7.3.9   Design of the filter bottom

It is important that backwash water be distributed over the entire filter bottom as uniformly as possible, so that all parts of the bed are washed equally. Many types of filter bottoms can be designed to accomplish this objective (see design calculations in Chapter 9).

Manifold and laterals

This is the oldest system. It has been used extensively throughout the world, and works well with proper hydraulic designs. A series of parallel, perforated pipes are inserted into a central channel or header. The size and spacing of pipes and perforations depend on the size and hydraulic requirements of the filter. Originally, most installations were made with a steel or concrete central channel or manifold with laterals made of steel, cast iron or asbestos cement. Today most of these filter bottoms are made of plastic piping (see Figures 7.1 and 7.2).

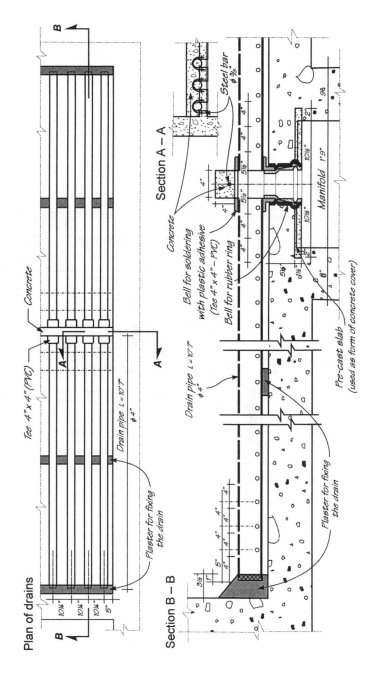

Figure 7.2 Design drawing for manifold and laterals in the filter bottom drains of a treatment plant for urban water-supply. Note the measurements are given in feet and inches.

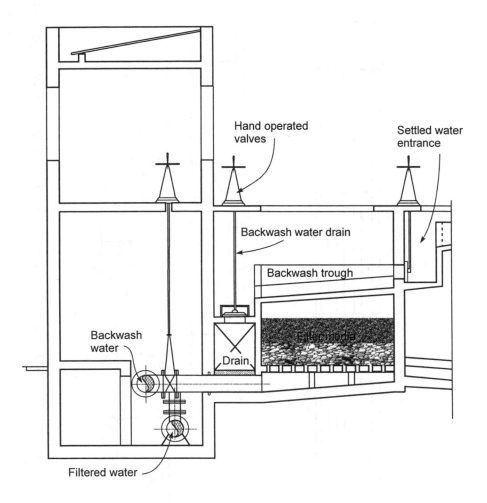

Figure 7.3 Existing plant filter and piping

A new version of this bottom is the corrugated type that was first used in California. In this design, the laterals are side by side and have a triangular or trapezoidal section (Figures 7.3–7.7). The central header with this arrangement of the laterals has excellent hydraulic characteristics and provides good filter washing. The laterals are held in place by a precast beam, which lies across the top of them.

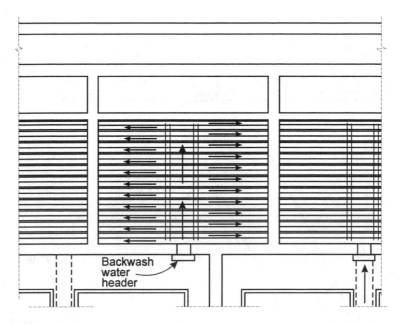

Figure 7.4 Plan view of corrugated filter bottom and backwash water flow

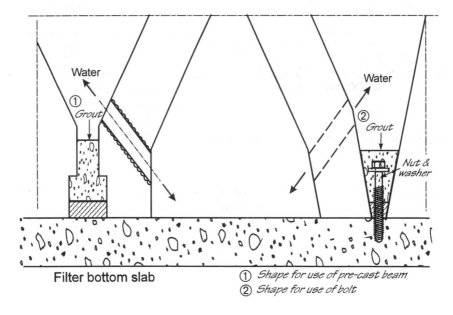

Figure 7.5 Corrugated filter bottom with only a small capacity for water

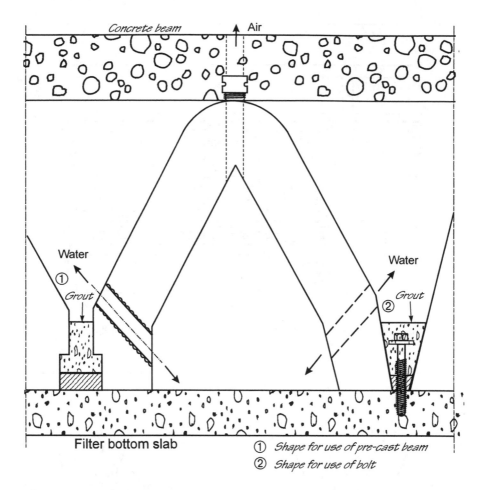

Figure 7.6 Corrugated filter bottom with air and water wash

Precast perforated blocks

The Leopold bottom is a dual manifold plastic block. It has excellent hydraulic characteristics, providing for both air and water wash. It is widely used in the USA. Plastic blocks are put together with rubber rings (see Figure 7.8).

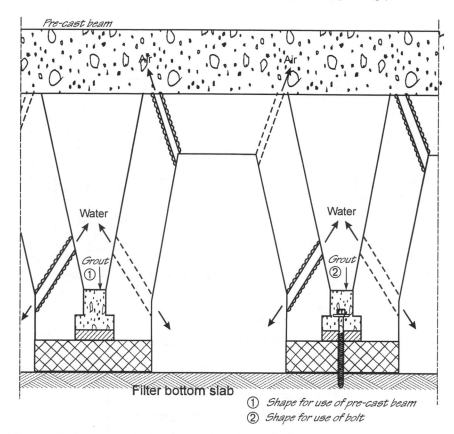

Figure 7.7   Corrugated filter bottom designed to provide a large capacity for water

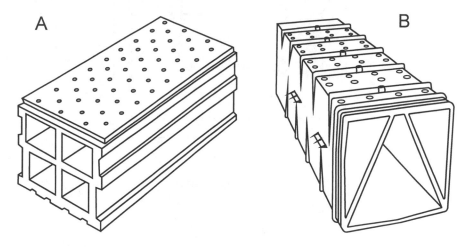

Figure 7.8   The Leopold filter bottom: dual manifold precast blocks

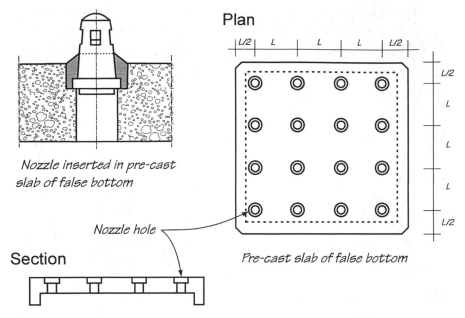

Plan

Nozzle inserted in pre-cast slab of false bottom

Nozzle hole

Section

Pre-cast slab of false bottom

Figure 7.9 False filter bottom of precast slabs with nozzles for backwashing

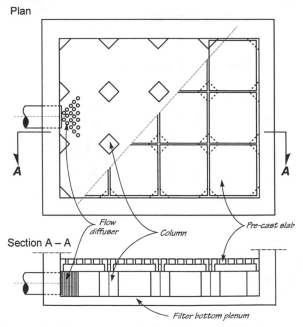

Plan

Flow diffuser

Column

Pre-cast slab

Section A – A

Filter bottom plenum

Figure 7.10 A precast slab arrangement with plenum and diffuser

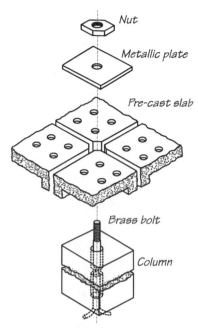

Figure 7.11   Detail for securing precast slab

False bottoms with nozzles

This consists of a concrete slab built over a plenum in the bottom of the filter box. Nozzles are inserted in precast slabs that make up the false floor. The nozzle manufacturer provides the specifications for hydraulics, air, and for spacing (see Figure 7.9–7.11).

Chapter 8

OPTIMISATION AND UPGRADING OF A PLANT
FROM 100 l s^{-1} TO 250 l s^{-1}

8.1 Assessment of existing plant

The design flow of the existing plant was 100 l s^{-1} but it was being operated at 125–150 l s^{-1} and the treated water was of poor quality. The plant layout (Figure 8.1) is a typical horizontal flow treatment with a Parshall flume for measuring flow.

8.1.1 Layout and dimensions

Alum is applied above the flume at one point as indicated, and lime is applied at the same point during the rainy season. From the flume, the water goes to a five compartment Alabama-type flocculator. The total volume is 183.75 m^3, which gives a theoretical flocculation time of 30.63 minutes. Although this time is ample, there is much dead space and therefore short circuiting occurs in this type of flocculator.

From the flocculator, the water flows to three settling basins by way of a manifold 1.00 m wide by 0.80 m deep, entering each basin through two gates 0.30 m wide and 0.70 m deep. The three settling basins are 6.75 m wide and 23.70 m long. The depth varies from 3.60 m at the entrance to 4.20 m at the sludge drain and up to 3.00 m at the outlet. There is a permeable baffle at the entrance with four rows of 0.10 m square ports. Each row of each settling tank has eight parts. Settled water is collected in three channels 3.75 m long, located one at each side of the basin and one in the middle as shown in Figure 8.1.

The settled water flows directly to five filters, which are 3.75 m by 3.95 m each. The sand depth in the filters is 0.60 m and the support gravel 0.45 m. The filters have false bottoms with nozzles spaced 0.20 m on centres. Under the false bottom is a 0.40 m deep plenum. Backwash water is collected in troughs that empty into a drain gullet 0.80 m wide. The mud valve drain is 0.45 m in diameter. The settled water enters the filters through sluice gates 0.30 m square and is distributed onto the beds through the wash water troughs.

The filters have independent pipelines (Figures 8.2) for filtered water (length 200 m) and backwash water (length 350 m). Filtered water flows to a baffled clear well (6.00 m long, 8.00 m wide, 2.50 m deep) under the chemical building (Figure 8.3). Chlorine is dosed at the entrance and the baffling

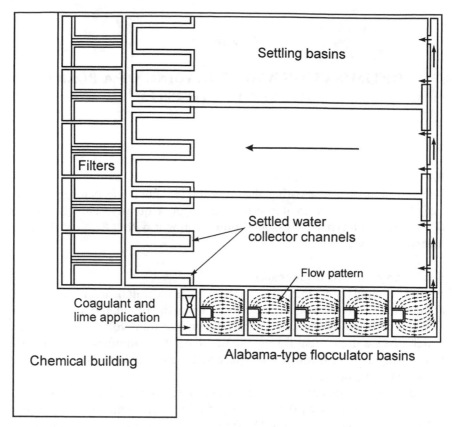

Figure 8.1 Unmodified layout of treatment plant, capacity 100 l s^{-1}

assures a minimum contact time. Lime is applied at the outlet for pH correction, then water flows by gravity to the city distribution system. Backwash water is pumped from the clear well to an elevated tank 10 m above the lip of the backwash overflow trough. The capacity of the tank (85 m^3) is sufficient for about 9.5 minutes wash at 0.60 m^3 m^{-2} per min.

The chemical and administration building has ample space for chemical preparation and storage, a laboratory, small office, baths, workers quarters, and the pump for elevating the backwash water.

The raw water varies somewhat between rainy and dry seasons, although an upstream dam alleviates sudden changes. The characteristics of the raw water are summarised in Table 8.1.

8.1.2   Plant performance
The raw water flow was being measured with reasonable accuracy but from there onwards the process control was very poor.

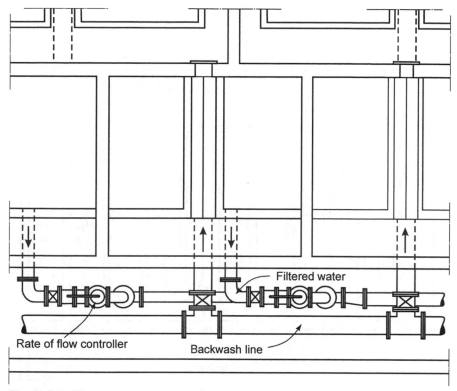

Figure 8.2   Filter piping in the unmodified treatment plant

Table 8.1   Characteristics of raw water during the rainy and dry seasons for a treatment plant with an original capacity of 100 l s^{-1}

	Rainy season			Dry season		
	Max.	Min.	Average	Max.	Min.	Average
Turbidity (NTU)	110	75	87	57	15	21
Colour (TCU)	23	15	17	16	4	6
Iron (mg l^{-1})	0.2			0.57		
Total coliforms (MPN)	3,200			100		
Faecal coliforms[1]	20			15		
Alkalinity (mg l^{-1})			15			30
Chloride (mg l^{-1})			35			55
Algae	Some problems in spring					

[1]   Number of coliforms per 100 ml   MPN   Most probable number

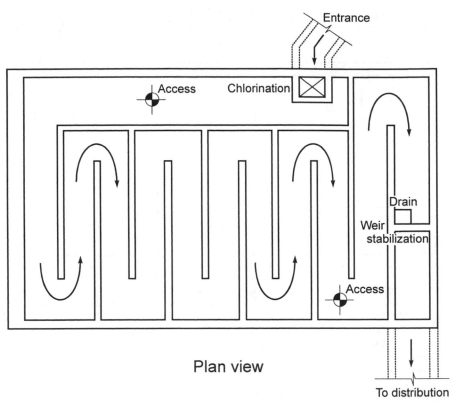

Figure 8.3 Plan view of a clear well

The alum $Al_2(SO_4)_3$ coagulant solution was applied as made up from solid, without any dilution. It passed through constant head feeders that were out of adjustment so that the exact dose was not always applied, nor was the applied dose always known. The coagulant was applied in one stream into the area just above the flume, which caused overdosing of a small portion of the flow and underdosing of a large portion. The Alabama type flocculators perform poorly because of the large amount of dead space and resultant short-circuiting. Clarification of samples dosed and flocculated in the laboratory was significantly better than achieved in the plant. This was also true of samples taken from the flocculator outlet and given further mixing in the laboratory, which indicates under-flocculation in the existing units. Samples from settling basins further confirmed the problem of poor coagulant dispersion and deficient flocculation.

Combined settled water turbidity was always 20–25 NTU, which is too high for the filters to perform well. The entrance baffle did not distribute the water evenly across the basin, and samples collected in the middle of the basin (10–12 NTU) were much better than those at the outlet. The sludge deposit in the settling basin was almost level, showing that far too much floc was being carried to the outlet end of the basin. Figure 8.4 shows typical configurations of sludge deposit for both well and poorly pretreated water.

Filtered water samples reflected the poor pretreatment shown by high turbidity of settled water. Filtered water turbidity was always 2.5–4.0 NTU during the period from 2 to 24 hours of the filter run. At the beginning it was higher, and from 24–36 hours of the run the turbidity increased to 12 NTU, which is far in excess of WHO guidelines for drinking water quality. Long filter runs were the result of control by head loss rather than filtered water turbidity.

Observations of the filters indicated their poor condition. Soundings showed movement of gravel and the mud ball tests gave results of 1.6–2.3 per cent, which is very high. The level of sand had dropped by 0.15–0.18 m in all beds and most of it had been deposited in the clear well. This was the result of damaged filter bottoms and broken nozzles and contributed to the poor quality filtered water and long filter runs. The most serious problem, however, was in the pretreatment. Most filters perform well if they are provided with properly clarified water.

8.1.3   Upgrading parameters determined by jar testing and experience

Procedures for jar tests are described in Chapter 5. With clear differences known between raw water characteristics of dry and rainy seasons, parameters were optimised separately for each.

Alum was found to give the best performance as coagulant, with dosage in dry and rainy seasons of 4–16 mg l^{-1} and 16–20 mg l^{-1} respectively. Lime was only required during the rainy season, dosed at 5–7 mg l^{-1}. Initial dispersion of the dilute alum solution was adequate at the weirs, with $G = 1,000$ s^{-1}.

Flocculation and settling parameters are shown in Table 8.2, optimised for both dry and rainy seasons although they were found to be very similar in both situations. The criterion for the settling process was to obtain clarified water with a maximum of 5 NTU.

8.2   Characteristics of the upgraded plant

Based on plant layout and the parameters determined in tests, a programme of modifications and expansions was compiled for the treatment of 250 l s^{-1}, or 2.5 times the existing nominal capacity (see Figure 8.5).

- *Coagulation.* Better conditions were needed for rapid and homogeneous mixing of raw water with coagulant.
- *Flow meter.* The existing Parshall flume was too small to measure the upgraded capacity of 250 l s^{-1} and therefore a splitter box had to be installed upstream for dividing flow accurately over sharp-crested, aerated weirs. Coagulant was applied at the weirs to encourage proper mixing.

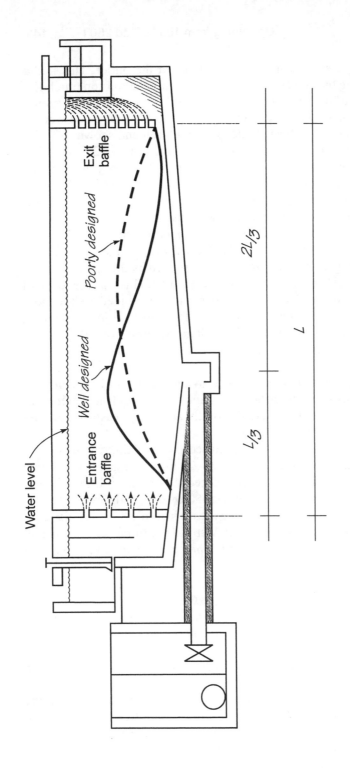

Figure 8.4 Longitudinal profile of sludge deposits in the settling basin, typical of well-flocculated or poorly-flocculated water

Table 8.2a   Optimum flocculation parameters (time and energy input) as determined by jar testing

Dry season		Rainy season	
Time (minutes)	Velocity gradient (s^{-1})	Time (minutes)	Velocity gradient (s^{-1})
2–3	50–60	2–3	50–60
3–5	35–45	3–5	35–45
5–7	25–35	5–7	25–35
7–12	15–20	7–12	15–20
Total not less than 22 minutes		Total not less than 20 minutes	

Table 8.2b   Optimum settling parameters (values for obtaining settled water turbidity not greater than 5 NTU) as determined by jar testing

	Dry season	Rainy season
Settling velocity (cm/min)	≤ 3.00	≤ 2.75
Surface loading (m^3 m^{-2}/day)	≤ 43.2	≤ 39.6

- *Flocculation.* By reducing the freeboard, the volume of the existing was increased to 192.94 m^3 but the volume for a minimum detention time of 25 minutes is 375 m^3 and therefore a new flocculator with at least 181.30 m^3 useful capacity had to be added. The old and new flocculators had to be fed in parallel to use the available head. If they were in series, the sum of the head losses would have been more than the available head. Baffles were needed in the existing flocculator for better flocculation and to control short-circuiting.
- *Flocculated water channel.* For equal flow distribution among the settling basins and to avoid floc breakage a new channel had to be built with three outlets. The channel was tapered to provide approximately the same flow velocity throughout, whilst water was discharged through the lateral outlets.
- *Settling tank.* The total area of the existing units for effective settling was 479.925 m^2 and 545.4 m^2 would be necessary for the surface loading of 39.6 m^3 m^{-2} per day (2.75 cm per min). Because the mean horizontal velocity would be about nine times the settling velocity of the lightest floc to be removed ($24.69 = 2.75 \times 8.97$), plate settlers for horizontal flow could be used for upgrading the tanks.

    The plate settlers had to be put near the outlet, while using the initial part for removal of the heaviest flocs. Note that the mean horizontal

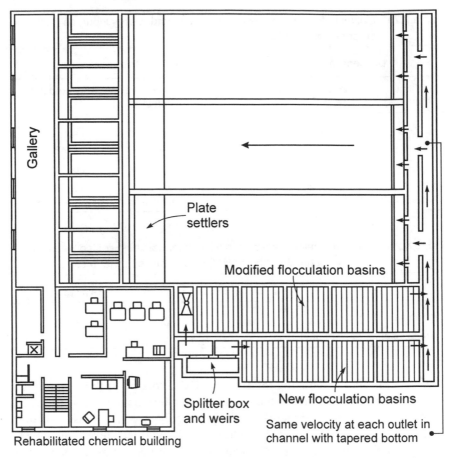

Figure 8.5 Layout of optimised and upgraded treatment plant, capacity 250 l s^{-1}. This modified design is contained largely within the existing plant structure

velocity for plate settlers with horizontal flow should be no more than 10–15 times the floc settling velocity.

- *Filter.* The existing area for filtration was 74.063 m^2 and therefore for the discharge to be treated (0.25 m^3 s^{-1}) the average rate was 12.15 m per hour. This was low for dual-media filtration and only a little high for sand alone, and thus new filters were not required. Entry sluice gates and filtered water pipelines had to be made larger for the increased discharge, and a new supplementary means for backwashing had to be used. Air scouring was introduced for auxiliary cleaning of the filter media. Filter sand and gravel had to be removed, washed and regraded. The gravel support layer was laid in reverse gradation to resist movement. Underdrains were repaired where necessary. The rate controllers were removed and the filters operated in the declining rate mode, resulting in better water quality and longer runs.

8.3   Inlet chamber

By splitting raw water flow as described above, the Parshall flume could still be used for measurement. Division of the flow was designed for the capacities of the subsequent flocculation basins. Because the existing basin was somewhat larger than the new one, the new basin should only receive approximately 52 per cent (192.9/375) of the flow (Figure 8.6). An approximate formula for calculating discharge over such weirs is given below:

$$Q = 1.8381 \, l \, h^{3/2}$$

Where Q = discharge (m^3 s^{-1})
        l = weir length (m)
        h = depth over weir crest (m)

The depth of the water over each weir should be around 10 cm, because this thin sheet of water is shallow enough for dilute coagulant to thoroughly mix with the raw water; if depth over the weir is too great then penetration is likely to be poor. The upgraded design flow was 0.25 m^3 s^{-1}, and the formula confirms that a total weir length of 3.55 m would be acceptable.

Calculated depth over the crest was a little over 10 cm but was satisfactory; and a proportional discharge to the flocculation basins could be achieved with weir lengths of 1.85 m and 1.70 m.

8.4   Initial mixing of coagulant and raw water

Deficiencies caused by a poor start to the water treatment process cannot be recovered, so it is worth considerable care in providing optimum conditions. Mixing of coagulant and raw water should be complete and very rapid, so that intended conditions for flocculation are provided throughout. The velocity gradient at the point of mixing should be about 1,000–1,200 s^{-1}, which is roughly equivalent to a fall of 10–15 cm over a weir (Figures 8.7–8.9).

The diffuser should be placed such that dilute coagulant solution (in this case 0.5 per cent alum) will impinge on the raw water a fraction of a second before the end of its fall. There should be at least 30 cm of freefall into the raw water, which will provide good penetration by the coagulant stream. The coagulant is applied in a thin stream along the length of the weir.

The raw water volume was divided into a thin, shallow stream passing over two weir sections. The coagulant was also applied all along the weir length as a very dilute solution and the raw water and coagulant were mixed at a point of high agitation. Therefore, everything was done to achieve almost instantaneous and total dispersion of all the coagulant with all the raw water as a first

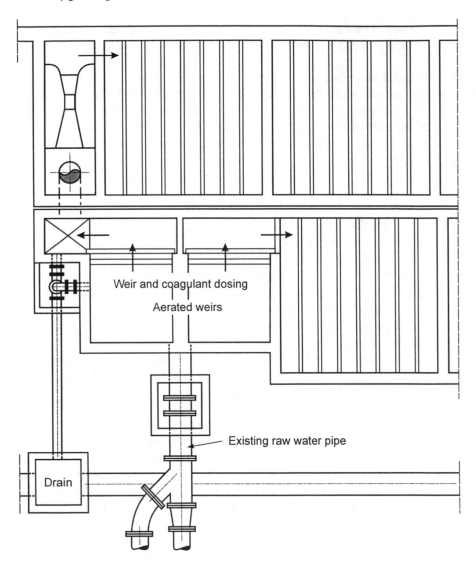

Figure 8.6 New splitter box and weirs, with coagulant dosing at the weirs for rapid dispersion

and vital step in pretreatment. A schematic view of the alum preparation and feed system is shown in Figure 8.10.

8.5 Flocculation

One of the problems of hydraulic mixing is the reduction in water level due to head loss as the water flows through the basin. This reflects energy used in

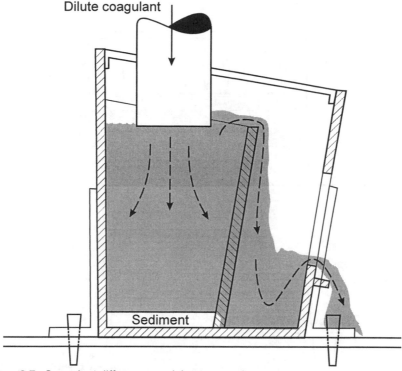

Figure 8.7 Coagulant diffuser comprising two sections

mixing. Water traversing the basin is agitated as it goes around the ends of the baffles. The applied energy depends on the velocity and area. If both are changing the process is difficult to control. Therefore, the conventional design of the baffling has to be modified (by keeping all of the openings fully submerged) so that only velocity changes and the area of flow around baffle ends is always fixed. In this way the velocity gradient is always known and under control (see Figures 8.11 and 8.12).

In Figure 8.11 the difference in the two designs is that, although the area of openings in the two designs is equal at the beginning, area A' will not change as water level changes. For a given flow, the velocity is related to the flow section and, because the velocity gradient is always directly related to the velocity, it is always known and can be controlled.

The design of the hydraulic flocculation system first required a division of flow between the old and new basins. This was controlled by the splitter box upstream which divided the flow into 0.13 m^3 s^{-1} in the old basin and 0.12 m^3 s^{-1} in the new basin.

The volume of the old basin was 193 m^3 (3.5 × 3.5 × 3.15 m × 5) through which the flocculation time with 0.13 m^3 s^{-1} flow was 24.74, or about 25

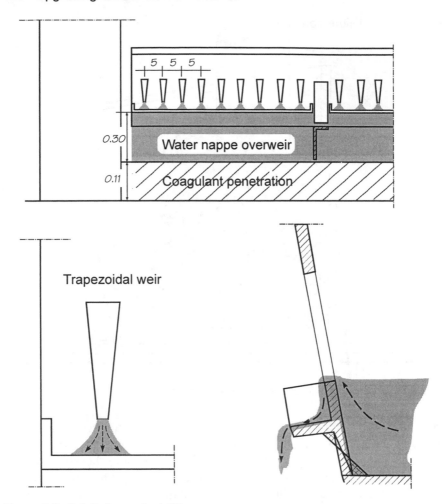

0.30

Water nappe overweir

0.11

Coagulant penetration

5   5   5

Trapezoidal weir

Figure 8.8 Detail of coagulant diffuser

minutes. An appropriate flocculation regime, determined by jar testing, was 3, 5, 5 and 12 minutes at 55, 40, 30 and 20 s^{-1} respectively.

The spacing of baffles in the basins (about 0.50 m) was such that workers could get into the basin and work between baffles. The number of spaces between baffles was 35 (3.5 m/0.5 m in each of five chambers) and the number of baffles was also 35, considering the chamber walls as baffles.

The residence time in each chamber was 4.94 minutes (24.74 min/5 chambers) and the time between baffles was therefore 0.705 minutes per space. The number of passages through baffles for the four stages of the optimum flocculation regime was therefore 4 (3 minutes at G 55 s^{-1}), 7 (5 minutes at G 40 s^{-1}), 7 (5 minutes at G 30 s^{-1}) and 17 (3 minutes at G 20 s^{-1}).

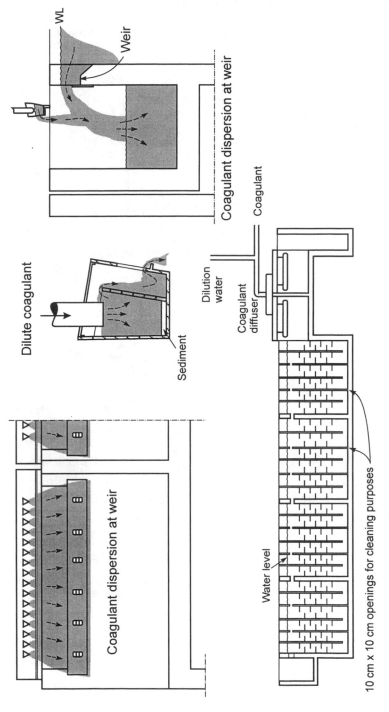

Coagulant dispersion at weir

WL

Weir

Dilute coagulant

Sediment

Coagulant dispersion at weir

Dilution water

Coagulant diffuser

Coagulant

Water level

10 cm x 10 cm openings for cleaning purposes

Figure 8.9 Coagulant diffuser and flocculator

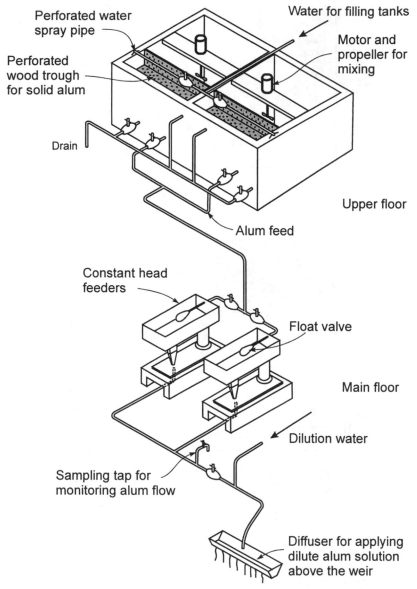

Figure 8.10 System for the preparation and feeding of alum solution

The area of baffle openings in the baffles can be calculated for each velocity gradient, by manipulation of two equations:

$$Q = Av \qquad \text{(i)}$$

and

$$G = 1.1334v^{1.50} \qquad \text{(ii)}$$

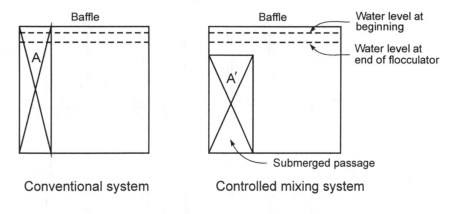

Conventional system   Controlled mixing system

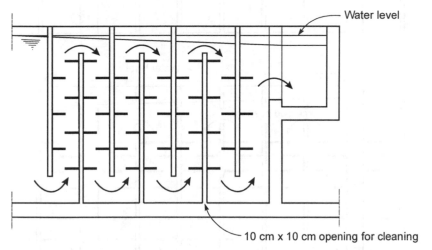

Over and under hydraulic flocculation conventional system

Figure 8.11 Hydraulic flocculation: water level and passage area in conventional and controlled systems; and layout of baffles in a conventional (over and under) flocculator

Where Q = discharge (0.130 m^3 s^{-1})
         A = area of opening (m^2)
         v = velocity (m s^{-1}) obtained by solving equation (ii)

Thus: $$A = 0.13 / (G/1334)^{2/3}$$

In this case for G of 55, 40, 30, 20 s^{-1} there needed to be four baffle openings of 0.99 m^2, seven of 1.13 m^2, seven of 1.48 m^2 and 17 of 1.94 m^2. There are openings in the baffles at the bottom to facilitate cleaning of the basin.

Plan

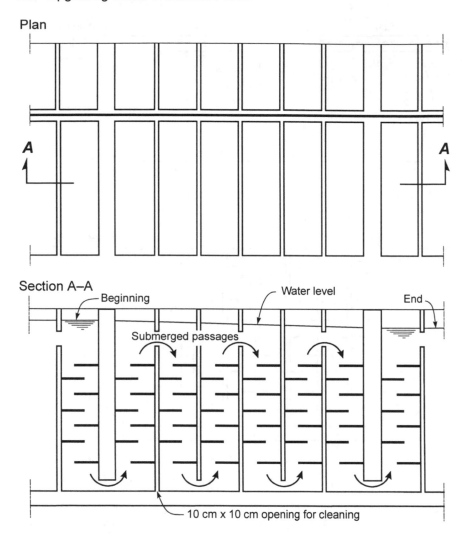

Section A–A

Beginning

Water level

End

Submerged passages

10 cm x 10 cm opening for cleaning

Figure 8.12 Hydraulic flocculation: layout of baffles in a controlled (over and under) flocculator, with the upper passages submerged

The methodology for the design of the baffling system for the new flocculation basin followed exactly the same procedure, with details adjusted for the slightly different dimensions.

8.6 Distribution of flocculated water to settling basins

From a common channel, water needed to be divided equally between the three settling basins, which hydraulically is a distributing manifold problem.

Effort in good design of the manifold is worthwhile (see section 5.6.3). There were several factors to be considered in the design of the channel:

- The floc that had been carefully formed in the flocculation basin must not be broken through high turbulence in the channel or in the ports.
- Velocity in the channel must be constant through its length for equal flow distribution.
- There were only three exit ports — one for each settling basin.
- The existing channel could be maintained as it was by limiting the depth of the last exit of the new manifold to that of the existing channel.
- Exit ports of the new channel were located so that discharging water did not impinge directly on any of the existing ports into the settling basins (see Figure 8.5).
- All the exit ports had to have the same dimensions.
- Flow velocity in the channel had to be sufficient to prevent the floc from settling, but at the same time not high enough to cause excessive turbulence that would break the floc.

To meet these requirements both hydraulics calculations and experience were needed and it was best therefore to address the theoretical basis for the design first of all.

Depths in the distributing channel
It was necessary to have a channel in which velocity and dimensions assured equal discharge from the exit gates of the manifold into each settling basin. The coefficient of head loss through the gates was proportional to $(V_m/V_L)^2$ where V_m and V_L are velocities in the manifold and in each exit gate or lateral (this relationship can be plotted as "Hudson's graph"). To provide equal discharge through gates, V_m/V_L has to be the same. Because the manifold discharge decreases after each gate and the size of all gates is the same, the only solution for this with a fixed width was to taper the bottom of the manifold channel, calculating appropriate depths at each gate. A means to find the relationship of depth to discharge is derived below.

The velocity in the channel is a function of the hydraulic radius, the slope of the channel and the coefficient of roughness of its surface, according to the Manning formula:

$$v = \left(R^{2/3} I^{1/2}\right)/n$$

Where v = velocity (m s^{-1})
         R = hydraulic radius (m)
         I = slope
         n = roughness factor (0.013 for concrete)

Two other expressions regarding power input are available:

$$\text{Power} = \mu V G^2 \text{ and Power} = QH\gamma$$

Where μ = dynamic viscosity (1.029 × 10–4 kgf × s m^{-2} at 20 °C)
          G = velocity gradient (20 s^{-1})
          V = volume (m^3)
          Q = discharge (m^3 s^{-1})
          H = head (m)
          γ = specific weight (998.23 kgf m^{-3} at 20 °C)

Combining these three expressions which cover flow, velocity gradient, head loss, velocity and size of the channels, the result for channel velocity is:

$$v = \left(\mu/\gamma\right)\left(G^2/I\right) = \left(\mu G^2/\gamma n^2\right)^{1/3} R^{4/9}$$

Which for rectangular flow section of width b and height h is:

$$v = \left(\mu G^2/\gamma n^2\right)^{1/3}\left[bh/(b+2h)\right]^{4/9}$$

For values of μ, γ, and n given already, and $G = 20$ s^{-1} selected by experience of successful designs, the relationship of v to b and h is:

$$v = 0.6249\left[bh/(b+2h)\right]^{4/9} \qquad\qquad (i)$$

Discharge Q is related to velocity and area ($Q = vA$ and $A = b\ h$) such that combining with (i) above:

$$Q = vbh \implies v = 0.6249\left[bh/(b+2h)\right]^{4/9} = Q/bh$$

If channel width is fixed (and velocity gradient is constant at 20 s^{-1} as substituted previously) this equation can be rearranged to find the required height for each discharge:

$$h\left[bh/(b+2h)\right]^{4/9} = Q/0.6249\ b$$

For width 0.80 m and three exits (one for each settling tank) the calculated depths before the first and last exits are 0.89 m and 0.35 m respectively. The former is satisfactory but the latter is too shallow because the upper end of the existing channel was 0.60 m deep. To accommodate this depth at the upper end, about 0.25 m had to be added to depths, so that the depth upstream of the first exit would be 1.15 m. This means that G in the channel will be less than 20 s^{-1}.

Dimensions of openings
To avoid floc breakage, openings in the new channel and the size of gates (two per basin) in the existing structure had to be determined for a velocity gradient of 20 s^{-1}. With one basin out of service, the flow through each gate would be 0.0625 m^3 s^{-1} (0.25 m^3 s^{-1}/4).

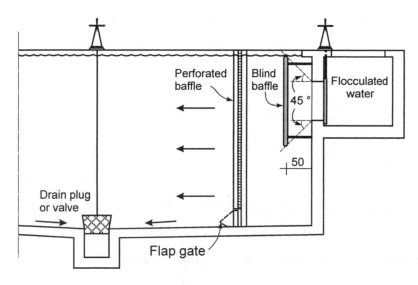

Figure 8.13 Location of baffles at the entrance to the settling basin

With a flow of 0.0625 m^3 s^{-1}, velocity was 0.26 m s^{-1}, when corrected for 20 °C. The gate area should be 0.24 m^2 (0.0625/0.26) and with a depth of 0.60 m the required width was 0.40 m.

8.7   Settling basin
Stages to consider in design were:
- The entrance of water, particularly entrance baffles.
- Effective settling of floc, with reference to settling velocity and residence time, using plate settlers if necessary.
- Removal of sludge from settling basins.
- Removal of settled water.

8.7.1   Entrance baffles
To avoid preferential currents, the flocculated water should not impinge directly onto the perforated baffle and therefore a stilling baffle was placed in front of the entrance gate to absorb kinetic energy and distribute the water across the area of the entrance. This had to be not less than 0.50 m from the inlet gate of the settling basin and much larger than the inlet gate determined by a projection of 45° from the sides and bottom of the exit gate (Figure 8.13).

The perforated baffle has many ports across its area (Figure 8.14), the number depending on flow and basin size. Basin entrances and entrance baffles in this plant were each 6.75 m wide and 3 m deep and each received a third (0.0833 m^3 s^{-1}) of the total flow.

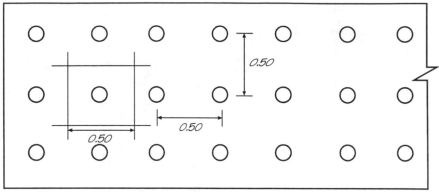

Ports: 100 mm diameter
Spacing: 0.50 m centre to centre

Figure 8.14 Perforated baffle for the settling basin entrance

Ports should be about 0.40–0.50 m apart, spread across the upper two thirds to three quarters of the baffle. Fewer, larger, widely-spaced ports or more, smaller, tightly-spaced ports can be used. The ideal is the largest number of ports that meet practical and hydraulic requirements. Ports are expensive to install but if there are too few they may be too widely-spaced.

There must be a head loss through the perforated baffle so that it is difficult for the water to short circuit. Water will always take the easiest, shortest course. Stilling baffles in front of the entrance gates absorb kinetic energy and disturb velocity paths, so that in seeking an easy access and not finding it, the water automatically spreads across the permeable wall.

One other condition should be met. The velocity through the ports should not be so high as to cause breakage of the flocs, but should provide a velocity gradient about the same as that in the final part of flocculation cycle, which in this case was 20 s^{-1}.

The rules for the design of the permeable or perforated wall can be summarised as follows:

- There should be the highest possible head loss through the ports, compatible with the G value.
- Velocity gradient through ports should be compatible with that at the end of flocculation.
- There should be the maximum practical number of ports, with the centres spaced at 0.5 m at most. This reduces the distance before jets merge and effective settling begins.

For practical reasons the ports through the perforated baffle should be made of short pieces of pipe of appropriate diameter to satisfy the design requirements. All material used must be able to resist the acid conditions of the settling basin. Plastics are the first choice for both economy and resistance to corrosion.

The inlet baffle should fit the settling basin for which it is being designed, which in this case was 6.75 m wide and 3 m deep at the entrance. A trial set of specifications can be compiled for assessment in relation to the general requirements.

- Port size: 75 mm, spacing 0.50 m between port centres.
- Port area: $(0.075 / 2)^2 = 0.0044$ m^2.
- Total port area: 0.0044 m$^2 \times 60 = 0.264$ m^2.
- Velocity: 0.0833 m^3 s^{-1} 0.264 m$^2 = 0.316$ m s^{-1}.
- Velocity gradient: about 35 s^{-1} (from Hudson graph, see section 8.6).
- Head loss: $(v2/2g) \times 1.7 = (0.3162/19.62) \times 1.7 = 8.7$ mm.

This trial for port size and spacing is adequate for all specifications except the velocity gradient, which is a little high (25–30 s^{-1} would be better). If the floc in this particular water was quite robust there would be no problem, but if the floc was weak (in coloured, soft, or cold waters) there would be a danger of floc damage. To reduce velocity, a pipe size of 100 mm can be tried.

- Port size: 100 mm, spacing 0.50 m between port centres (5 rows of 11 ports).
- Port area: $(0.100/2)^2 = 0.007854$ m^2.
- Total port area: 0.007854 m$^2 \times 55 = 0.432$ m^2.
- Velocity: 0.0833 m^3 s^{-1} $/ 0.432$ m$^2 = 0.1928$ m s^{-1}.
- Velocity gradient: about 22 s^{-1}.
- Head loss: $(0.1928/19.62) \times 1.7 = 3.22$ mm.

The baffle for this basin will function best by using 100 mm (4 inch) pipe for the ports, which would provide a more acceptable velocity gradient. The head loss would be a little low, although it should be good for overloads. To increase head loss the number of ports could be reduced to about 45–50, with an adjusted geometric layout of the ports. A head loss of about 5 mm would be better for distribution of flocculated water across the basin section. Velocity and velocity gradient would increase, but a gradient of 30 s^{-1} would be tolerable.

With the installation of the perforated baffle, the water would begin its passage through the basin in plug flow fashion. Velocities across the basin would be almost the same and floc particles would have the maximum chance to settle from the water column.

8.7.2   Settling performance and plate filters

The mean horizontal velocity for the design flow of 0.25 m^3 s^{-1} was 0.24 m per min with a clean basin and 0.30 m per min with the section reduced by 20 per cent due to sludge accumulation.

It is necessary to remove floc with a 2.75 cm per min settling velocity (decided in jar tests) and this amounts to 39.6 m^3 m^{-2} per day. The required surface area for this removal is 545.5 m², but only 479.9 m² was available. The difference between these two values is not great and therefore the existing basins could probably reduce the turbidity sufficiently to enable the filters to perform well without quickly clogging. The existing basins were able to remove floc with a settling velocity of 3.13 cm per min or more but to provide a margin of safety, plate settlers were installed in the outlet end of the basin. This would reduce the settled water turbidity to a very low level and allow for an overload without risk of lowering water quality.

The basic conditions for the plate settler design were:

- Removal of particles with a settling velocity of 2.75 cm per min or more.
- Horizontal flow which was not more than ten times the minimum settling velocity.
- Flow and short-circuiting control by a perforated outlet baffle with a high head loss (2–3 cm).
- The horizontal plates at 60° for sloughing of settled sludge from the plates.
- Distance between plates 5–30 cm (7.5 cm for this plant) with shorter plate settlers for more closely-spaced plates. Efficiency is independent of plate spacing, which depends on available space, cost and facilities of assembling and cleaning.

Because the plates are set at an angle of 60° to the horizontal and are 7.5 cm apart, the maximum settling distance would be 15 cm (7.5/cos 60°).

The settling velocity of the floc to be removed was 2.75 cm per min and, because it has at most 15 cm to fall, 5.45 minutes (15 cm/2.75 cm per min) had to be allowed within the apparatus. The mean horizontal velocity with the chosen design flow was 24–30 cm per min, which was adequate to meet the criterion that horizontal velocity would be within ten times the settling velocity (i.e. 27.5 cm per min). The latter could conveniently be used to calculate an appropriate plate length, namely 1.5 m (27.5 cm per min/5.45 min). In fact, to allow for turbulence on entry to the settler, the plates had to be 2 m long (see Figures 8.15–8.18).

A more correct and elegant method of calculating the plate length is the Yao formula:

$$(V_s/V_o)(\sin\theta + L\cos\theta) = S_c$$

Where V_s = settling velocity of floc to be removed (m per min, in this case 0.0275)

V_o = average flow velocity (m per min). In this case the design velocity for the basin was 0.232 m per min, being $(0.250 \text{ m}^3 \text{ s}^{-1} \times 60 \text{ s})/(3 \times 6.75 \text{ m} \times 3.20 \text{ m})$, but if 10 per cent of the basin section was allowed for the settler, $0.258 \text{ m}^3 \text{ s}^{-1}$ was likely to be more accurate

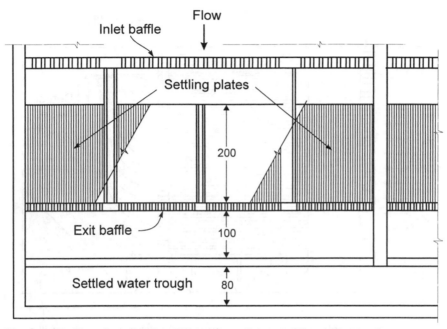

Figure 8.15 Plan view of plate settler at the outlet end of the settling basin

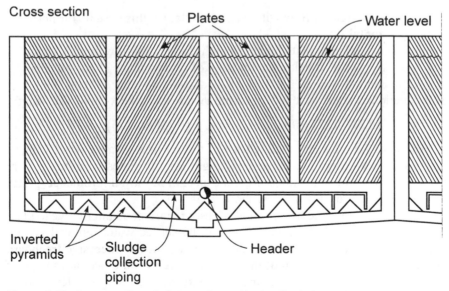

Figure 8.16 Cross section of plate settler and hydraulic sludge collection and removal system

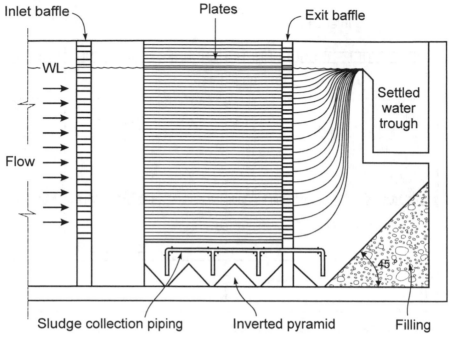

Figure 8.17   Longitudinal section of settling plates

θ = angle of flow with the horizontal (in this case 0° for horizontal flow)

L = relative plate length (length divided by flow depth)

S_c = critical value for equation of performance (1 for parallel plates)

This equation can be rearranged to provide a value for L, the relative plate length:

$$(0.275/0.258)(0+L)=1 \quad \Rightarrow \quad L=(0.258/0.0275)=9.37$$

For plates 7.5 cm apart, inclined at 60°, the settling distance was 15 cm, as discussed previously. The calculated length for the plates was thus 1.41 m (9.37 × 15 cm) and, again allowing some margin for initial turbulence, plates of 2 m length were confirmed as appropriate.

If the water department were short of resources, a good alternative to plate settlers would have been improvement of the settled water removal system and raising of the water level to take advantage of the excessive freeboard (see Figure 8.19). Extension of the collecting weirs could take advantage of clearer water in the centre of the basin and raising the water level could reduce horizontal velocity. Under most conditions these small changes would provide settled water of good quality. Nevertheless, plates would provide more security, but at a somewhat higher cost.

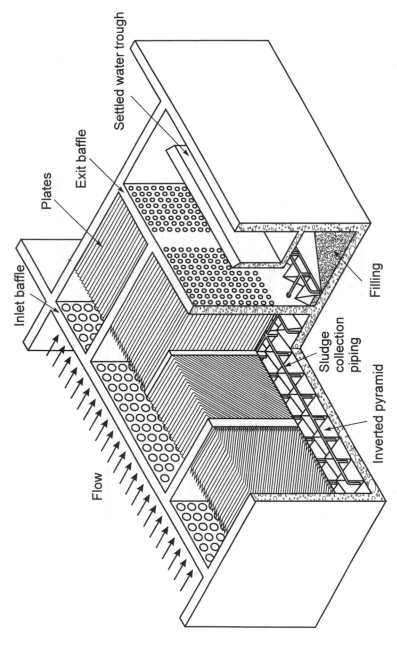

Figure 8.18 Isometric view of settling plates

Settled water trough

Exit baffle

Plates

Inlet baffle

Flow

Sludge
collection
piping

Filling

Inverted pyramid

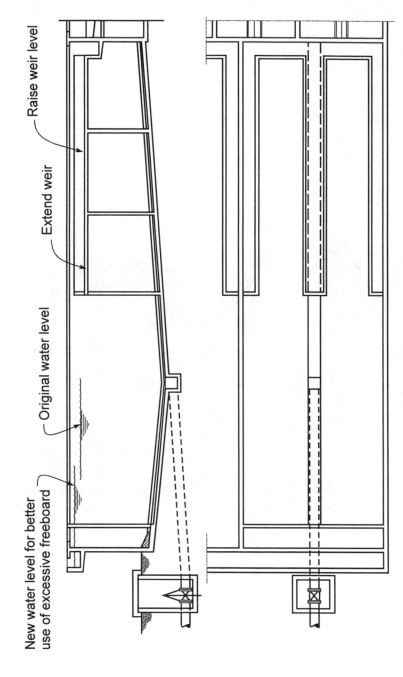

Figure 8.19 An alternative upgrading option: water level raised and outlet weirs extended

8.7.3 Sludge removal system

Floc settles out on the plates and slides down their inclined surfaces to the bottom or to the 0.20 m corridor and accumulates in the basin bottom beneath the plates, where a hydraulic system is constructed for periodic sludge removal.

The frequency of sludge removal (by opening the valve for about one minute) depends on the amount of solids in the raw water, which varies during the year. Automatic valves with timers can be installed and set for a specific removal cycle. The initial costs and maintenance of these are much more expensive than manually operated valves, but are often the choice if labour is costly. Figures 8.15–8.18 show the layout of the plates and the sludge removal system. The controlling factors for the design of the sludge removal system were:

- Volume of sludge accumulating beneath the plate settlers.
- Minimum head loss per opening of 0.15 m. Head loss in the lateral opening should be five times that in the lateral, which in turn should be three times that in the header.
- Space available beneath the settling plates and arrangement of inverted pyramids.
- Minimum size of lateral pipe was 38 mm (1.5 inches).
- Small openings could be made by attaching a cap to the lateral with a hole drilled out.

The volume of the sludge-collecting inverted pyramids (in this case 0.112 m^3) can be found using the following formula:

$$\left(h/3\right)\left(B+b+\sqrt{Bb}\right) = 0.112 \text{ m}^3$$

Where h = height (0.42 m)

         B, b = longer and shorter base sides (0.84 m, 0.10 m)

The sizes of the sludge-collecting orifices in the lateral, the lateral itself and the header, have to be calculated as indicated below.

Size of the sludge-collecting orifices in the lateral

$$Q = C_v s \sqrt{2gh}$$

Where Q = discharge, which for one pyramid was 0.0019 m^3 s^{-1}
(0.112 m^3/60 s)

         h = head loss through orifice (m)

         g = acceleration due to gravity (9.81 m s^{-2})

         C_v = coefficient of discharge (0.65)

         A = flow section (m^2)

         $A = 0.0019/\sqrt{0.65 \ \ 2 \times 9.81 \times 0.15} = 0.0017$ m^2

         D = diameter (m) of orifice

         $D = \sqrt{4A/\pi} = 0.0466$ m or 1.75 inches

For this removal system the sectional area of each orifice should be 0.0017 m², corresponding to an orifice diameter of 0.0466 m (1.75 inches).

Size of the lateral
Diameter of the lateral can be calculated by using a rearrangement of the Hazen-Williams equation:

$$h_f = \left(Q^{1.85}L\right)\bigg/\left[\left(0.2785C\right)^{1.85}\times D^{4.87}\right]$$

Where h_f = head loss (0.03 m, being 0.2 of that in sludge-collecting orifice)
Q = discharge (which was 0.0038 m³ s⁻¹ for two pyramids)
L = length (1.85 m)
C = 120 for plastic pipe
D = diameter of lateral (m)

$$D = \left(Q^{1.85}\times1\right)\bigg/\left[\left(0.2785\times120\right)^{1.85}\times0.01^{1/4.87}\right] = 0.07 \text{ m}$$

Substituting parameters above, an appropriate diameter for the lateral was 0.074 m (3 inches).

Size of the header
The size of the header can be calculated as above. Head loss (0.01 m) was one-third of that in the lateral, discharge from 12 laterals was 0.0456 m³ s⁻¹, and header length was 1.68 m. The result gives a header diameter of 0.234 m (about 10 inches).

8.7.4   Removal of settled water
Removal of the settled water was through another perforated baffle at the outlet. This had a much higher head loss than the entrance baffle. During early settling, the water is filled with floc which has been carefully formed and great care must be taken not to cause it to break, but once most of the particles (especially larger ones) have been removed, there is less need to be concerned about floc break-up. Stresses applied in the filter are also much higher than any encountered during passage between the settling and filtration units. The exit baffle can therefore be designed with a head loss of about 2 cm, which controls the flow in the settler so that no currents or short-circuiting occur which would reduce efficiency of that system.

The three baffles each had 86 holes, so each hole had to convey 0.001 m³ s⁻¹ under the design flow of 0.250 m³ s⁻¹. Velocity, and diameter of the holes, could therefore be calculated as follows:

$$v = \sqrt{2gh_f/k} = \sqrt{(2\times9.81\times0.02)/1.7} = 0.48 \text{ ms}^{-1}$$
$$D = \sqrt{4A/\pi} = 0.051 \text{ m}$$

Where $A = Q/v = 0.001/0.48 = 0.0021 \text{ m}^2$

To increase head loss a little, 80 holes distributed over the settled surface of the baffle were satisfactory. Because the baffle only serves to distribute flow through head loss, pipe inserts were not needed. An economical and effective baffle could be made of wood with drilled holes.

The effluent of the settling basin was collected in the existing channel for conveyance to the filters. This channel was able to manage the increased flow. The increased velocity did not cause a problem because the water had been settled and almost all the floc had been removed.

8.8   Filters

For the filters to perform well under increased flow conditions, the media had to be changed from existing units to local coal over sand. Ideally, media specifications should be based on pilot filters, otherwise they can be estimated from experience and from information in the literature.

At the top surface of the filter there was a layer of coal, of depth 0.45 m, effective size 1.2 mm, and uniformity coefficient of not more than 1.25. The sand layer immediately beneath the coal was 0.25 m deep, with an effective size of 0.65 mm and a uniformity coefficient of not more than 1.15. Both sand and coal were fairly coarse but very uniform.

Part of the sand came from the existing filters after washing, cleaning, and regrading. The coal was obtained from national sources to avoid the high cost of imported anthracite, although if anthracite is available it should preferably be used. Most bituminous coals (specific gravity 1.49–1.52) are rather lighter, softer and less resistant than anthracite (s.g. 1.57–1.60), although they still perform well and replacement may not require importation. Many Latin American countries, that have been using local coal for many years, have found it very satisfactory.

The filter support layer of gravel also had to be modified. Instead of a conventional gradation of gravel size from large at the bottom to small at the top, the size was increased again at the top. This reverse gradation helped resist gravel movement which is a common problem in filters. Details of the profile are given in Table 8.3 and are illustrated in Figure 8.20.

The false bottoms made of concrete with nozzles were retained, except for repairs that were found necessary when the filters were emptied for rehabilitation. Some nozzles were broken and needed replacement.

The filtered water outlet piping had to be enlarged to accommodate the augmented design flow, and consequent filter rate of 12.5 m^3 m^{-2} per hour (maximum about 18 m^3 m^{-2} per hour). The outlet line was 175 mm (7 inches) diameter, which had to be increased to 300 mm (12 inches) and the new filtered water collector line (Figure 8.21) had to be 400 mm and 500 mm.

Table 8.3   Profile (size, depth) of support gravel in filter beds

	Gravel size		Depth	
	mm	inches	cm	inches
Top	25–50	1–2	12.5	5
	12.5–25	$^1/_2$–1	5	2
	6.3–12.5	$^1/_4$–$^1/_2$	5	2
	3.1–6.3	$^1/_8$–$^1/_4$	10	4
	6.3–12.5	$^1/_4$–$^1/_2$	5	2
	12.5–25	$^1/_2$–1	5	2
Bottom	25–50	1–2	10	4

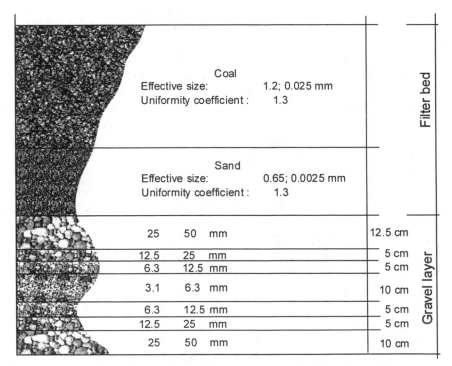

Figure 8.20 Profile of filter media and the support gravel which is in partial reverse gradation

Filter control had to be changed to declining rate, the existing rate of flow controllers had to be removed and all the filters had to be controlled from a

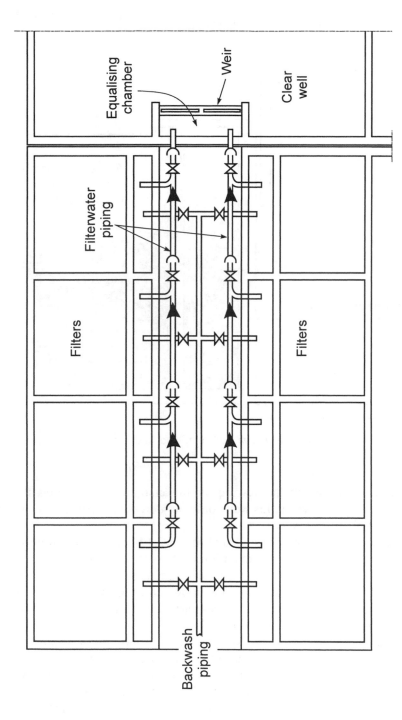

Figure 8.21 Filtered water and backwash water piping, equalising chamber and clear well

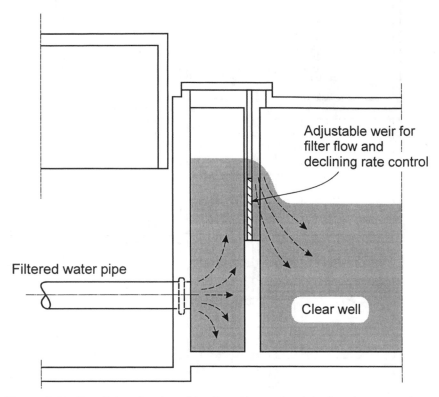

Figure 8.22 Equalising chamber with adjustable-level weir for filter flow (declining rate control)

variable level weir in the new equalising chamber (see Figures 8.21 and 8.22). The water level at this weir could be increased to reduce the available head and plant flow rate.

A perforated disc was installed at the outlet of each filter, to introduce a head loss for control of maximum filter rate when the filter was clean. Eight or nine holes were required to control surge amplitude by minimising flow fluctuations. At any instant, the fluctuations through all the small diameter holes would not be of the same magnitude and direction, thus smoothing pulses more effectively than a single hole.

Filter rate was controlled hydraulically with no valves, venturi or other devices, because these would have tended to worsen water quality (because of flow fluctuations) and because they are expensive to acquire and maintain in proper working order. In the declining rate control system, the available head is limited so that as the filter clogs, head loss increases and the filter rate declines. Ultimately filtration stops and, with no further head available, the filter overflows. When operated well, the filters are washed long before total clogging occurs, but this is a valuable protective feature because poor back-washing leads only to a waste of water, rather than to filter damage.

With an equalising chamber and a variable-level weir the filter could not over-produce or create negative pressure, even when clean, and therefore an outlet disc should not have been required (although it would have provided an added safety factor). In many plants, the clear well is directly below the filters, and a perforated disc must be installed, unless it is feasible and practical to combine the filter flow into a chamber. Design of both the equalising chamber weir and perforated disc is discussed in Chapter 9, which describes upgrading of a 1,000 l s^{-1} treatment plant.

Figure 8.23 shows a double-disc plate that can be adjusted to create exactly the desired head loss by controlled occlusion of five trapezoidal openings.

8.9   Summary

This plant was in poor condition, operating above its design capacity to treat 125–150 l s^{-1} of water with precarious final quality. The optimised and upgraded physical plant is now in good condition and provides 250–270 l s^{-1} of water which is of excellent quality. The costs of the additional capacity have been estimated at 30 per cent of a corresponding new construction.

Mixing of coagulant with the raw water has been improved to provide rapid and complete dispersion by installation of two new weirs at a splitter box and proper coagulant dilution to 0.5 per cent (rather than 20 per cent). Flow is measured at the existing Parshall flume.

A new hydraulic flocculation basin has been designed using the technique of submerged openings, so that the necessary velocity gradient can be applied under complete control. The old one has been modified to eliminate short circuiting and provide the correct G values. Bench scale pre-design testing was used to provide the necessary information for correct mixing along the length of the flocculation basin.

The entrance to the settling basin has a perforated baffle to distribute flow equally starting as a plug flow at uniform velocity across the basin. To maintain clarification under adverse conditions, a 2-metre unit of plate settlers (with hydraulic sludge removal) has been installed to reduce the minimum settled floc size as determined in pre-design testing. The upstream and downstream parts have thus been improved for settling of both heavy and light flocs, so that water reaching the filters will be of the lowest turbidity that can practically be provided.

The filters have been rehabilitated after constant over-loading with turbidity for years. The control system has been changed to declining rate in which one weir controls the rate on all filters and the rate-of-flow controllers have been removed. Some new piping was necessary to accommodate the doubled design flow. Filter bottoms have been retained, with support gravel

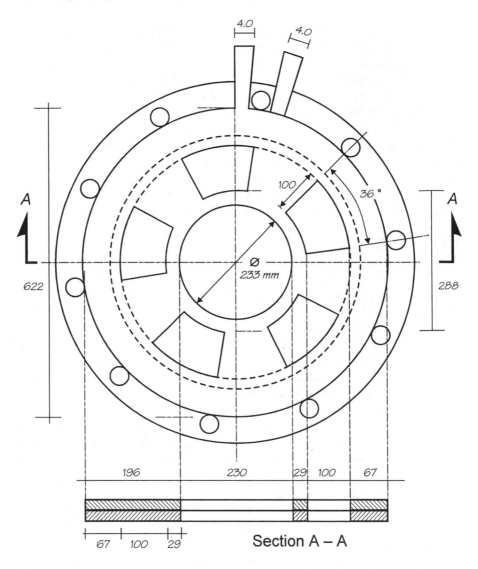

Figure 8.23 Plate with adjustable trapezoidal openings, for control of the maximum filter production (dimensions appropriate for 200 l s^{-1} discharge)

laid in reverse gradation, and with dual media filter beds of sand and locally-sourced coal. The chemical building and operational facilities have been renovated so that operators have a pleasant working environment. Chemical preparation and feed has been redesigned so that the process is easy to operate and control. In reality, a new 250 l s^{-1} plant has been created, largely within the existing structures at a very low cost.

Chapter 9

OPTIMISATION AND UPGRADING OF A PLANT
FROM 1 m^3 s^{-1} TO 2.5 m^3 s^{-1}

9.1 Assessment of existing plant

This plant was designed and constructed for production of 1 m^3 s^{-1} (86,400 m^3 per day). It had been operating overloaded by 20–30 per cent for many years and the treated water was of poor quality. Expected demand in the future was 2.5 m^3 s^{-1}. The treated water quality needed to be improved because for much of the time it exceeded the WHO turbidity guideline of 5 NTU.

9.1.1 Layout and dimensions

Figure 9.1 shows the layout of the original plant. Raw water entered through a venturi of 1.3–1.4 m^3 s^{-1} capacity (which was in poor condition and would need replacing) and was then directed to a rapid mix basin of 2.75 m × 2.75 m × 4.00 m which provided about 30 seconds of detention. Mixing was provided by a mechanical turbine and electric motor of 25 horsepower. Concentrated coagulant was applied at one point in the corner of the mixing basin. Rotation of the turbine was too slow for good dispersion. Therefore much of the water was underdosed and a small amount was overdosed.

From the mixing basin, water flowed to a flocculation system of six chambers (8.5 m square, 3.5 m deep) each with four mechanical, rotary paddle agitators. There was a great deal of short-circuiting and dead space, otherwise the flocculation time of 25 minutes would have been sufficient.

The manifold distributing dosed water among the flocculation basins was not designed for dividing the flow equally because the importance of this was not known at the time. This channel needed to be redesigned and modified accordingly. Water flowed directly from the flocculation system to the settling basins. Each of the three basins (17.25 m wide, 54 m long, average depth 4.55 m) received water from two flocculation chambers. Distribution of flocculated water across the settling basins was rather poor and thus a new permeable baffle had to be installed at the entrance to each basin. Sludge was removed with mechanical scrapers that were in reasonably good condition. Settled water passed from each basin over a weir that was the width of the basin at the outlet end. The overflow rate was high and swept up a large

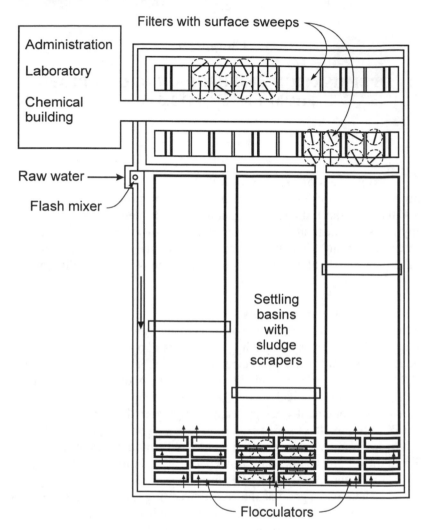

Figure 9.1   Unmodified layout of treatment plant, capacity 1,000 l s^{-1}

amount of floc. Water was collected by a channel at the end of the basins for transport to the filters as shown in Figure 9.1. The plant was designed for (and was operating with) a freeboard of 0.45 m which was more than necessary and could probably be used more efficiently in the modifications for upgrading.

There were 12 filters in two banks of six with a gallery between each bank. Each filter was itself divided into two beds of 3.75 m × 8.20 m, separated by a central gullet. Sand of 0.70 m depth overlay support gravel, which was 0.50 m deep. The underdrain system consisted of a central main channel with perforated laterals of cast iron pipe. Each side of the filter was backwashed

Table 9.1   Characteristics of raw water during the rainy and dry seasons for a treatment plant with an original capacity of 1,000 l s^{-1}

	Rainy season (4 months)			Dry season (8 months)		
	Max.	Min.	Average	Max.	Min.	Average
Turbidity (NTU)	500	25	48	25	15	18
Colour (TCU)	50	20	25	15	10	12
Alkalinity (mg l^{-1})			25			80

separately using water that was supplied by pumping. Rotary sweeps provided auxiliary sand cleaning during the backwash but many were not functioning. The filters each had a rate of flow controller that did not function properly.

9.1.2   Plant records and performance

Plant records were sparse but some information was available. The raw water was pumped from a large river. The characteristics of the river (Table 9.1) differed considerably between dry and rainy seasons.

The poor plant performance started with initial mixing of the coagulant which was piped at full preparation strength (about 20 per cent) through constant head feeders to the mixing basin. A stream of coagulant was applied in one corner of the basin where ineffective mixing and the small volume of concentrated coagulant accounted to a large extent for the poor quality of settled water.

The dosed raw water was transported by a long channel, which was not designed to provide equal distribution between the flocculation basins. The entrance baffles were not designed to distribute flocculated water across the basin section with the result that flow was uneven and some of the water traversed the basin very quickly. The basins had one compartment with four paddle-type mixers and this configuration further tended to increase the proportion of water that short-circuited. The result of these deficiencies was a poorly-formed floc and incomplete settling. High settled water turbidity was encouraged by the outlet weir, which lay simply across the width of the basin. This was the shortest practical weir design, at which the overflow rate was highest. The consequent upsweeping current carried a large amount of floc with it.

With pretreatment, which was inefficient throughout, the high turbidity of settled water did not allow good filter performance. Output from the filters was of variable quality but filtered water turbidity exceeded the WHO guideline value of 5 NTU for most of the time.

The chemical building, control office, and laboratory were in poor condition and needed to be completely rehabilitated.

9.1.3   Upgrading parameters determined by jar testing and experience

Based on bench scale testing and experience of factors that influenced plant performance, the following design and process parameters were determined:

- The best coagulant was ferric chloride ($FeCl_3$) dosed in the range 3–35 mg l^{-1}.
- All chemical preparation, storage and feed equipment had to be operated in line with the required coagulant volume.
- Polymers were not required for good treatment. The range in which they might have contributed was considered to be uneconomical.
- Initial mixing of coagulant and raw water needed to be done with G not less than 1,000 s^{-1}.
- Flocculation time was 30 minutes. Tapered energy input during flocculation was best with 60–80 s^{-1} for 5 minutes, 30–40 s^{-1} for 5 minutes, then 15–20 s^{-1} for 20 minutes.
- Surface loading of settling basins should be at 3.2 cm per min (46 m^3 m^{-2} per day) for 4–5 NTU output.
- Dual media filters of sand and local coal were needed with an average filter rate of 12.5 m per hour. Reverse gradation of support gravel would be required over about 0.50 m depth. There should be a backwash rate of 0.80 m^3 m^{-2} per min. Filter control should be by declining rate.
- With the increased flow, some gates, piping, channels and ports would have to be changed to avoid high turbulence and head loss. This requirement had to be studied unit by unit.
- Treatment by direct filtration was possible in the dry season, when 3–5 mg l^{-1} of coagulant would destabilise the colloids enough to achieve turbidity below 1 NTU by filtration alone.

9.2   Modifications required for upgrading and optimisation

Pumping and pipeline
To increase flow from 1.0 m^3 s^{-1} to 2.5 m^3 s^{-1}, more pumps had to be installed in the modified pumping station. Three new pumps each of 1.25 m^3 s^{-1} capacity were installed, allowing one as a standby. Existing pumps were rehabilitated to provide further standby capacity. Construction of a new pipeline (150 cm diameter) was required to take more water to the plant.

Flow measurement and initial mixing of coagulant
A 10 m long sharp-crested, aerated weir (divided into five equal sections of 2 m each to provide good aeration) was used to measure flow and for

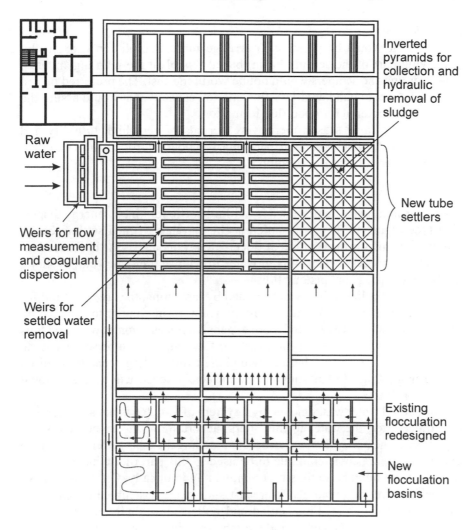

Figure 9.2 Layout of optimised and upgraded treatment plant, capacity 2,500 l s^{-1}.
This modified design is contained mainly within the existing plant structure

dispersion of the dilute coagulant. The weir, coagulant diffuser and their
arrangement are shown in Figure 9.2 and in Figures 8.7–8.9. Management of
flow into the plant and mixing of coagulant with raw water were improved,
initiating treatment in the best possible manner. In particular, both flow and
dosage were more closely measured and controlled and a more thorough
initial dispersion of the dilute coagulant was achieved.

Flocculation
Normally, it would be possible to arrange more flocculation capacity by modifying part of the settling basin, but at this plant the rearrangement of sludge drains and scrapers would not be worth the cost. The preferred option was, therefore, to build auxiliary flocculation capacity that would assure the proper time required for good floc formation.

During the rainy season there were periods of very high turbidity and surface runoff, which required particularly good control of the coagulant dosing, dispersion and flocculation.

Settling
A new perforated baffle preceded by a blind baffle for kinetic energy dispersion (see Figure 8.13) was installed at the entrance to each settling basin, initiating a uniform distribution of flocculated water across the basin section. The settling process began with all water flowing at the same velocity along the basin (a "plug" or "piston" flow).

Given the settling velocity determined by jar testing, the area required for simple settling could be obtained with two new basins similar to the existing ones. This would have required more expensive flocculation as well, and thus the preferred option was to install vertical flow, high-rate tube settlers at the outlet end of each basin. These provided the extra settling capacity required by the upgrading of plant output without costly construction of additional settling basins.

The settled water was removed by a system of V-notch weirs and channels that transported the water to the filters, according to the layout shown in Figure 9.2. The heavy floc settled before reaching the tube settlers, so that the existing scrapers could continue to be used in this part of the basin. Sludge from beneath the tube settlers was removed hydraulically. On the bottom of the basin, inverted pyramids of precast slabs were assembled as shown in Figures 8.16 and 8.18. Floc settled and collected as sludge in the bottom where it could be removed periodically by opening the valves on the sludge drain lines.

The inverted pyramids on the tank bottom were formed by longitudinal triangular prisms and inverted-V dihedral planes. There were five isosceles triangular prisms in the middle part of the basin and each side formed a rectangular prism, as can be seen in Figure 8.16.

Filters
The filter structures could be used without any change, but the content of the filter system had to be modified in line with the filter rate resulting from increased plant flow.

The modified filter media was sand from the existing beds (after it had been washed and cleaned) and above the sand was locally-sourced coal. The support gravel had a reverse gradation to resist movement. Instead of operating at a constant rate, the control was changed so that filters operated at a declining rate. The underdrain system had to be rebuilt to allow higher filter and backwash rates. The capacity of the wash water and filtered water lines also had to be increased.

9.3   Design of modifications

9.3.1   Chemical preparation
The same approach was used in this plant as discussed for the 250 l s^{-1} plant (Chapter 8). Chemical solutions were distributed to their points of application by gravity. The coagulant (ferric chloride) was supplied as a very acidic liquid and therefore the tanks had to be fully protected with acid-resistant paint and all piping had to be PVC.

The maximum expected coagulant dosage was 35 mg l^{-1}, and thus for a possible overload of 25 per cent, 9,450 kg per day had to be allowed for. The specific weight of commercial FeCl$_3$ was approximately 1.3 kgf l^{-1} and therefore supply and handling of about 7.27 m^3 per day may have been required.

The source of supply was about 50 km from the plant and the road was good, and therefore a 20-day stock held in storage tanks of total volume 145.4 m^3 (150 m^3) should have been enough. Two tanks of 75 m^3 were used for storage and supply, using corrosion-resistant materials throughout. For structural reasons the tanks were round, of 5 m inside diameter and 4 m useful depth. The bottom of each tank was sloped towards the outlet drain to facilitate cleaning. The stock was adequate to allow a tank to be taken out of service as required for repair and maintenance.

9.3.2   Application of coagulant
The coagulant flowed under gravity to the diffuser at the weir, where it was diluted to a 0.5 per cent solution for application. Ferric chloride was delivered to the plant at a concentration of about 30 per cent, and therefore about 59 parts of water were used to dilute the stock coagulant. The methods of measuring dilution water and mixing were as in the 250 l s^{-1} plant (Chapter 8).

The weir was 10 m long and divided into five 2-m sections as shown in Figure 9.2. The arrangement and location of the diffuser was exactly the same as in Figures 8.7–8.9.

9.3.3   Channel from inlet weir
Destabilised colloidal particles need to be kept in an environment of relatively high agitation to form strong, dense floc, and thus the velocity gradient

in the channel should be equal to or slightly higher than that in the first flocculation basin. In this case, it should not have been less than 80 s^{-1}. An expression for G can be derived by combining the two expressions:

$$P = \mu V G^2 = Qh\gamma \text{ and } v = \left(R_H^{2/3}I^{1/2}\right)\bigg/n$$

to give:

$$G = \sqrt{\frac{\gamma}{\mu n}}R_H^{3/4}I^{3/4}$$

For a channel of rectangular section (width B and depth H) the G value is:

$$G = \sqrt{\frac{\gamma}{\mu n}}\left(\frac{BH}{B+2H}\right)^{3/4}I^{3/4}$$

Taking B of the existing channel, which is known (1.5 m), and calling:

$$G\bigg/\left(\sqrt{\gamma/\mu n}\ I^{3/4}B^{3/4}\right) = A$$

the design variable H can be determined as:

$$H = \left(A^{4/3}B\right)\bigg/\left(12A^{4/3}\right)$$

The solution in this case was $H = 1.48$ m.

Depth of flow in the existing channel was 2.25 m and the freeboard in the settling tank was 0.45 m. The freeboard can be reduced to 0.20–0.25 m and this simple procedure provided the following clear advantages:

- Increased volume in the flocculation basins.
- Increased height in the settling basins for installation of the tube settler system.
- Increased water height over the filter beds.

With a freeboard of 0.20 m the water height within the channel had to be reduced to 1.65 m (1.45 m + 0.20 m). This was done by filling in the channel bottom as shown in Figure 9.3.

Mean flow velocity in the channel was 1.15 m s^{-1}, being 2.5 m^3 s^{-1}/(1.5 m × 1.45 m). The time required for water to go from the inlet chamber to the flocculation basins was about 54 seconds (62.2 m/1.15 m s^{-1}) and because the hydraulic grade for the flow was 0.00057 (from the Manning formula):

$$\text{Slope} = V^2 n^2\big/R^{4/3}$$

and variation in level over the 62.2 m was 0.036 m.

9.3.4   Channel for dividing flow among the flocculation basins

The main concern was to obtain equal flow distribution among the three flocculation basins, which was governed by the square of the ratio of the conduit velocity to the lateral velocity (Figure 8.5 shows distribution to three settling basins). Velocity in the lateral was constant, and thus velocity in the manifold had to be as constant as possible. Because flow in the manifold decreases

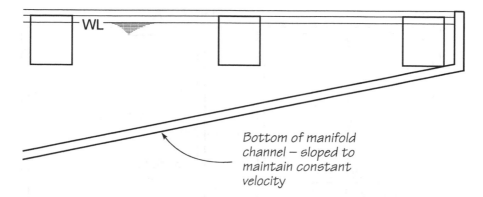

Figure 9.3 Tapered manifold bottom to maintain constant velocity with decreased discharge

after each exit to a flocculation basin, the flow section had to be reduced by tapering the manifold.

The entry ports of the three series of flocculation basins had to be designed for a velocity gradient compatible with that in the first flocculation compartment (in this case 60–80 s^{-1}) and at the same time a modest head loss. For G of 60 s^{-1}, the mean flow velocity was 0.412 m s^{-1} and head loss was about 0.9 cm. For a channel 1.2 m wide and G of 60 s^{-1} the depth was 2.25 m and mean velocity was about 0.926 m s$_{-1}$. After the first lateral outlet, flow was reduced by a third (to 1.667 m^3 s^{-1}). In order to maintain a constant velocity the cross-section of the manifold had to be 1.80 m^2 (1.2 m wide, 1.5 m deep). The flow after the second lateral outlet was again reduced to 0.833 m^3 s^{-1} and therefore the manifold cross-section had to be reduced to 0.90 m^2 (1.2 m wide, 0.75 m deep). A tapered manifold, which was 2.25 m deep at the first lateral outlet and 0.75 m deep at the third, provided good flow distribution among the three flocculation basins (see Figure 9.3).

9.3.5 Flocculation basins
The volume required for 25 minutes of flocculation at 2.5 m^3 s^{-1} flow was 3,750 m^3. Normally an additional 10 per cent was allowed for short-circuiting and dead space but in this plant there was almost two minutes agitation in transit from the entrance weirs to the flocculation basins; and with six compartments (Figure 9.2) the possibility for short-circuiting was much reduced.

The basins originally provided a total capacity of 1,625.62 m^3 (six basins, dimensions of each one 8.5 × 8.5 × 3.75 m). Much more flocculation capacity was needed, and so six new basins were installed as shown in Figure 9.2. The old basins were compartmentalised and together with the new basins provided a series of six compartments in each of the three sections of the

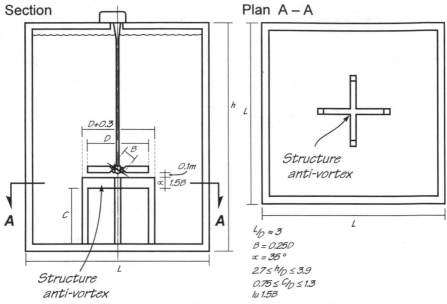

Section   Plan A – A

Structure
anti-vortex

$L/D \approx 3$
$B = 0.25D$
$\alpha = 35°$
$2.7 \leq h/D \leq 3.9$
$0.75 \leq C/D \leq 1.3$
$lu\ 1.5B$

Structure
anti-vortex

Figure 9.4 Mechanical flocculator with pitched propeller and anti-vortex structure

flocculation system. Additional capacity of the new basins was 2,254.2 m³ (six basins, each 8.5 × 8.5 × 5.2 m) for a total of 3,879.82 m³, which met the required 3,750 m³ with a small factor of safety.

Detention time in each new basin was about 7.5 minutes, which was rather longer than testing indicated. To compensate partially, intensity of agitation was reduced so that the input was 60–70 s⁻¹ in the first basin, 35–45 s⁻¹ in the second, and 15–20 s⁻¹ in the remaining four basins.

Mechanical agitation in the new basins was provided by propellers, each with four flat blades. The length of the propeller blades and the width were determined experimentally. The length of the propeller blade (2.5 m) was roughly one third of the length of the side of the chamber, while the 0.625 m propeller width was one quarter of its length. Figure 9.4 shows the arrangement and dimensions. The pitch of the propeller blades (at 35° to the horizontal) was also derived from hydraulic experiments in mixing basins.

The power to be applied to the water for a given velocity gradient can be simply calculated as:

$$P = \mu V G^2$$

Where μ = dynamic or absolute viscosity (1.029 × 10⁴ kgf s m⁻² at 20 °C)
   V = volume (8.5 × 8.5 × 5.2 = 375.7 m³)
   G = velocity gradient (e.g. 70 s⁻¹ in first basin and 45 s⁻¹ in the second basin)

On the other hand, for each type of fluid agitation in vessels, the analysis in terms of an energy equation is based on a relationship between the power number (N_p) and Reynolds number (N_{Re}), which depends on the type of agitator, the characteristics of the vessel, and the liquid.

$$N_p = Pgc/N^3D^5\rho \quad \text{and} \quad N_{Re} = D^2N/v$$

Where gc = dimensional constant or factor of Newton Law
  Kg m/(Kgf × s^2) = 9.8066 m s^{-2}
  N, D = rotational speed (rev s^{-1}), diameter (m) of propellers
  ρ = density of the fluid (101.7917 kgf m^{-4} s^2 for water at 20 °C)
  v = kinematic viscosity (1.007 × 10^{-6} m^2 s^{-1})

For a baffled system and N_{Re} not less than 10,000, experiments show that N_p is constant. With a propeller of 1 m diameter, rotating at 0.1 rev s^{-1}, the N_{Re} at 20 °C is:

$$N_{Re} = \left(1^2 \times 0.1\right) / \left(1.007 \times 10^{-6}\right) \approx 99{,}304 > 10{,}000$$

Because the propellor diameter and the rev s^{-1} are always, respectively, greater than 1 m and 0.1, the N_{Re} is larger than 10,000 and N_p is always constant. The experimentally-determined value for the propellers to be used in these flocculation basins was 1 and thus:

$$Pgc/N^3D^5\rho = \mu VG^2 gc/N^3D^5\rho = 1$$

Substituting the values given above in this equation gives:

$$1.026 \times 10^{-4} \times 8.5 \times 8.5 \times 5.2 \times 9.8066 \, G^2 = N^3 \times 2.5^2 \times 101.7917$$

Thus the relationship between N^3 and G^2, or N and $G^{2/3}$, enables the determination of the rotational speed of the propellor, N (in rpm), for any G value as indicated below:

$$G^2 = 241.8737 \, N^3 \quad \text{or} \quad N = 0.1605 \, G^{2/3} \quad \text{or} \quad N_{rpm} = 9.6299 \, G^{2/3}$$

The power to be applied determines the specification of the motors that are used to drive propellers at each stage of the tapered energy input. Again values for this plant can be substituted in the expression for power, to obtain P for a required G, as shown below:

$$P = \mu VG^2 = 1.021 \times 10^{-4} \times 8.5 \times 8.5 \times 52 \times G^2 = 0.0384 G^2$$

Substituting various values for G produces the corresponding values for the power P to be applied to the water as follows:

G (s^{-1})	P (kgm s^{-1})	HP
70	187.96	2.5
60	138.10	1.84
45	77.68	1.04
40	61.37	0.82
35	46.99	0.63
30	34.50	0.46
20	15.30	0.20
15	8.60	0.11

To take into account all the other factors influencing the total power required, such as starting torque and the friction in gears and bearings, theoretical power was increased 4–6 times. In other words, the motor specified for the first basin applying 70 s^{-1} would have to be rated at 10–15 HP and a 15 HP motor should probably be specified to ensure performance. This same rationale was applied to the motors in subsequent flocculation basins.

9.3.6   Settling basin

The settling basins had a total effective settling area of 2,794.5 m^2 (three basins, each 17.25 × 54 m) and the surface loading for the upgraded flow of 2.50 m^3 s^{-1} was 5.37 cm per min, which was too high for good clarification and for a reasonable margin of safety. The mean horizontal velocity of 0.604 m s^{-1} with the new flow was also very high.

One solution to the problem would have been simply to build more settling basins. To achieve the removal of floc with settling velocities of 3.2 cm per min, the area required was 4,687.5 m^2. Two new basins were needed to give the additional 1,893 m^2 basin area but it was much cheaper to install tube settlers in the existing basins for additional clarification capacity.

From experience, suitable settlers have 5 cm square tubes in which velocity is 15 cm per min. The tubes were constructed in blocks 0.75 m wide, 0.54 m deep and 3.0–3.6 m long. Each block was made up of a series of tubes in rows. Alternate rows sloped in opposite directions (making an X shape) and were fused so that they supported each other and the block was very strong. The tube walls were less than 1 mm and the structure remained relatively light for its strength.

Maximum settling distance in these tubes was 10 cm (5 cm/cos 60°) and the settling velocity of particles to be removed was 3.2 cm per min. The general equation for the flow trajectory of suspended particles in the tube is given below (the Yao equation), from which the average flow velocity required in the tube settler (V_o) was found to be 0.1653 m per min.

$$V_s\left(\sin\theta + L\cos\theta\right)/V_o = S_c$$

Where V_s = settling velocity of floc to be removed (0.032 m per min from jar testing)
θ = inclination of tubes to the horizontal (60°)
S_c = 11/8 (parameter for square tubes)
V_0 = velocity of flow in the tubes (using the same units as V_s)
L = relative length (l / d = 12.47) where length l = 0.54 m/sin 60°, width d = 0.05 m

The area of the settlers was 1,047.49 m^2, determined as total discharge of 150 m^3 per min (2.5 m^3 s^{-1}) divided by the vertical component of flow velocity (0.1653 sin 60° = 0.1432 m^3 per min). Because there are three basins, each had a flow area of 349.16 m^2 and allowing 10–12.5 per cent for settler tube walls, structure and space for assembly, the total required was 384–393 m^2.

Blocks of settler tubes were 3.44 m long and 0.75 m wide, and five blocks fitted into each basin with a 5 cm allowance for irregularities and convenience in placing blocks. With 30 rows of five blocks each there was a total settler area of 387 m^2 plus the allowances for irregularities and fitting the blocks in place, which was enough to meet the requirement.

With the tube settlers installed at the outlet end of the tank, there was 28.15 m of basin for plain settling to occur, and the heavier flocs settled out before reaching the tube settlers. The sludge removal equipment continued to operate in the first part of the basin.

A large quantity of the less heavy floc settled out under the tube section. This sludge was removed hydraulically after accumulating in the bottom of inverted pyramids. These pyramids were constructed of precast slabs and assembled in the bottom of the basin, as shown in Figures 8.16. This bottom design had two functions: to collect the sludge at the bottom of the inverted pyramid and to distribute flocculated water evenly throughout the area under the tubes. Three ridges were formed longitudinally along the bottom of the basin and outlet ports were built into the precast slabs to allow the flocculated water to discharge along the length of the triangular prismatic conduit. Velocities were quite low, and therefore no problems were anticipated, although there was some floc settling and once or twice yearly the entire bottom area had to be washed out.

Settled water was collected in precast concrete troughs above the tube settler sections. These collection troughs caused some disturbance of flow distribution in the settlers, but several recommendations were made to minimise their influence:

- The shape of the troughs had to be as hydrodynamic as possible.
- Lips of overflow weirs to the troughs had to be at least 1 m above the top of the settlers.

- The troughs had to be spaced horizontally not more than three times the vertical distance between the top of the tube settlers and the bottom of the troughs.
- Although they have a V-shaped bottom, the troughs were calculated for varied flow, with a rectangular lateral spillway section with a horizontal profile and no friction slope.

As shown in Figure 9.2, this plant had three settling basins with a central collecting channel and eight collecting troughs on each side, spaced according to the criteria stated above. The central channel was 1.0 m wide with walls 0.15 m thick and each transverse trough was 7.975 m long. The normal flow was 0.0521 m^3 s^{-1} and the expected maximum was 0.0781 m^3 s^{-1}, which would occur when one settling basin was out of service. The equation for the flow profile is:

$$(x/l)^2 = \left(1 + 1/2F^2\right) y \big/ y_o - 1/2F^2 (y/y_o)^3$$

Where x = distance from the origin (m)
y, y_o = water height at x and at the outlet (m)
l, b = length, width of the trough (m)
F = Froude number, $F^2 = ql^2 \big/ gb^2 y_o^3$
g = acceleration due to gravity, 9.8066 m s^{-2}
q = rate of inflow per unit length of trough (m^3 s^{-1} m^{-1})

If there is free fall from the lateral trough into the main collecting channel, the Froude number is 1, the flow profile is:

$$(x/l)^2 = 1.5(y/y_o) - 0.5(y/y_o)^3$$

When the outlet is submerged, the depth y_o is the water level at the outlet or the depth of the water at the outlet.

For troughs with the design flow of this plant, 0.25 m wide, the parameter F_2 as a function of y_o is:

$$F^2 = 0.0521^2 \big/ \left(9.8066 \times 0.25^2 \, y_o^3\right) = 0.0044 \big/ y_o^3$$

and thus the expression for the flow profile is:

$$(x/l)^2 = \left[1 + \left(y_o^3 / (2 \times 0.0044)\right)\right] y / y_o - y_o^3 \big/ (2 \times 0.0044) \times \left(y^3 / y_o^3\right)$$

At the uppermost end of the trough ($x = 0$) the latter expression can be simplified considerably to obtain the value of y with respect to y_o as follows:

$$y^2 = \left(y_o^3 + 2 \times 0.0044\right) \big/ y_o$$

For y_o of 0.20 m, y at the head of the trough was about 0.29 m, and therefore it was clear that the trough of 0.25 × 0.30 m would not be flooded by receiving maximum flow from the settling basin.

The trough section is shown in Figures 9.5 and 9.6. Additional capacity in the V section of the trough (beyond that recognised in calculations above)

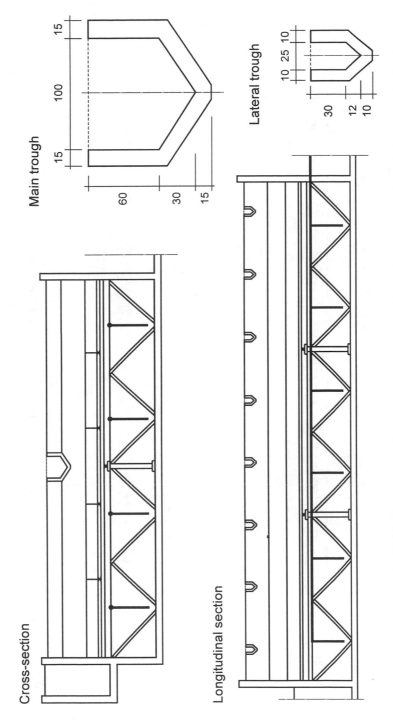

Figure 9.5 Settled water collection: longitudinal section through the lateral troughs and cross section through the main collecting trough

Hydraulic profile in the main channel

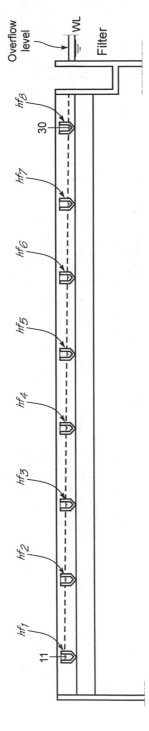

Hydraulic profile in the first lateral trough

Figure 9.6 Settled water collection: hydraulic profiles in the main channel and lateral troughs

provided a further guarantee against flooding, which would otherwise have completely distorted flow conditions in the tube settlers.

For the main collecting channel, the upper end of the hydraulic profile was the water level of the first or uppermost lateral from which it received discharge. As indicated above there was a 0.09 m head loss in the lateral, such that the outflow level was about 0.11 m below its overflow.

Two head losses need to be considered for the determination of the hydraulic profile of the main collecting channel: friction slope, which is determined by the Manning formula and is usually negligible, and flow confluence, given by the expression:

$$K\left(v^2/2g\right)$$

Where v = velocity in the channel downstream of discharge from the lateral (m s^{-1})

g = acceleration due to gravity, 9.81 m s^{-1}

K = coefficient depending on the discharges of the confluence

The value of K is:

$$K = 1 + \left(Qu/Qd\right)^2 - \left(Qu/Qd\right)^2 \left[\left[1+\left(Qu/Qd\right)\right] \middle/ \left[0.75+0.25\left(Qu/Qd\right)\right]^2\right]$$

Where Qu = flow in the main channel upstream of the confluence (m^3 s^{-1})

Qd = flow in the main channel downstream of the confluence (m^3 s^{-1})

The main collecting channel was 1.0 m wide and 0.60 m deep with a V-shaped bottom. At the upstream end, the water level was about 0.11 m below the top and the flow section was 0.74 m^2 which avoided flooding the transverse troughs. With one settling basin out of service, the flow in the lateral troughs was 0.0781 m^3 s^{-1}.

Velocity downstream of the uppermost confluence was 0.211 m s^{-1}, because a discharge of 0.1562 m^3 s^{-1} (2 × 0.0781) was passing through a 0.74 m^2 section. There was no flow above this first confluence and therefore $Qu = 0$; consequently $K = 1$ and the head loss h_f was:

$$h_f = (1 \times 0.211^2)/(2 \times 9.81) = 0.0023 \text{ m}$$

For the second or the next downstream confluence, the downstream discharge Qd was 4 × 0.00781 = 0.3125 m^3 s^{-1} and the upstream discharge Qu was 2 × 0.0781 = 0.1563 m^3 s^{-1}, therefore $Qu/Qd = 0.5$ and:

$$K = 1+(0.1563/0.3125^2)^2 - (0.1563/0.3125)^2 \, [[1+(0.1563/0.3125)] / [(0.75+0.25\times(0.1563/0.3125))]]^2$$
$$= 1 + 0.25 - 0.25 \times [(1+0.50) / (0.75+0.125)^2]$$

$$= 1.25 - 0.4898$$
$$= 0.76$$

Because velocity downstream of the confluence depends on the head loss, the upstream data was used for a first approximation, which was therefore $0.211 \times 2 = 0.422$ m s^{-1}. Head loss then was:

$$h_f = (0.76 \times 0.4222) / (2 \times 9.81) = 0.007 \text{ m}$$

The flow section and velocity can be found by summation of the two head losses already determined ($0.0023 + 0.007 = 0.01$ m) and used to refine the head loss estimate. Flow section was reduced to 0.73 m^2 and therefore the average flow velocity was 0.428 m s^{-1}. Replacement of this value in the above expression makes no significant difference to the estimate of head loss (0.007 m) and the water level in the main channel was therefore about 0.12 m below the top.

For the third confluence, using the same procedure, the water level was 0.132 m below the top because h_f is about 0.012 m. For the fourth confluence, h_f is about 0.017 m and the water level was 0.15 m below the top of the main channel. Continuing these calculations for the remaining confluences, water level downstream of the eighth or last confluence was 0.31 m below the top of the channel.

The relevance of this final water level to the avoidance of flooding (and thus proper function of the tube settlers) was not obvious without specific knowledge or experience. Flooding of troughs is often seen in filters during backwashing because the designer did not understand the hydraulic procedure to be applied to the backwash water drainage system. The result is poor filter wash, clogged filters and poor filtered water quality. Filter problems and water quality deterioration are sure to appear under this situation.

The head loss in this main channel was quite high. If it could not be accommodated, for plant flow reasons, the solution would have been to build two main collecting channels. This would have reduced head loss to about a quarter that of one channel (7–8 cm) but the construction costs would have increased.

The hydraulic sludge removal system was similar to that of the 250 l s^{-1} plant discussed in Chapter 8. Spacing and dimensions were different but the design procedure was exactly the same.

In tube or plate settlers, the sludge slides down the surfaces in clumps, because settled floc particles accumulate until the component of their combined mass weight in the direction of the settler is large enough to overcome the friction force, and as clumps slide down they further collect other deposited particles. At the same time as the larger clumps of sludge are settling to the bottom, the suspended flocs are rising, but the settling sludge traps some of the suspended floc and the effectiveness of clarification is increased. This action is similar to that of rising floc particles going through a sludge blanket in combined units.

9.3.7   Filters

In rehabilitating old filters or in designing new ones, it is necessary to consider first the backwash, which still is the highest flow and an important aspect of filter technology. Design of the backwash system is deficient in most old filters. When observing filter washing, it is common to see that one part of the filter washes more than another because of poor distribution of the backwash water. Another serious problem in older plants is that the backwash rate is too low.

The backwash rate for the rehabilitated filters was based on a flow of 0.80 m^3 m^{-2} per min and a wash cycle of 10 minutes. The two filtering beds of each filter were 30.75 m^2 (8.20 × 3.75 m). Thus to wash each filter would require 492 m^3 of backwash water and the two halves were washed in sequence, one immediately after the other.

The wash water piping had to be able to provide 25 m^3 per min (0.417 m^3 s^{-1}) for the total 20-minute wash and the velocity should not have exceeded 2.8 m s^{-1}. The necessary pipe size was therefore at least 0.1489 m^2 (0.417/2.8), equivalent to 0.435 m diameter. Piping of 450 mm diameter may have been available, but otherwise 500 mm diameter would have been suitable.

It might seem that 2.8 m s^{-1} would be a high velocity, and indeed it was, but flow was of quite short duration and the valves were never closed so rapidly that it caused a problem of water hammer.

With dual media filter beds, an auxiliary wash is needed to keep the bed in good condition. In this plant, sweeps were installed originally and they were retained, replacing any which were in poor repair. Water pressure required for proper operation was 55 m water at the nozzles. When the bed was filtering, the sweeps needed to be about 8 cm above the level of the filter media. During the backwash, the sweeps were operating within the coal layer of the expanded bed.

The size of the central gullet in these filters had to be adequate to receive 0.417 m^3 s^{-1} during the backwash. The discharge equation:

$$Q = 1376bh_o^{3/2}$$

can be rearranged for water height h_o.

Where Q = total flow (m^3 s^{-1})
         b = width of gullet (m)
         h_o = water height at the upstream end (m).

For a gullet of 0.75 m width and a total flow of 0.417 m^3 s^{-1}, the water height at the upstream end was calculated as follows using the rearranged equation:

$$h_o = (Q/1376b)^{2/3} = [0.417/(1376 \times 0.75)]^{2/3} = 0.55 \text{ m}$$

As the water height at the upper end of the gullet was 0.55 m, the depth at that point could be 0.65 m. The slope had to be about 5 per cent to give enough velocity to maintain all solids in suspension and keep the gullet clean.

The gate entrance to the main drain had to be large enough to avoid flooding of the troughs and gullet, which is a common cause of poor filter backwashing. Assuming the maximum depth over the gate to be 0.50 m, the gate area could be calculated as follows:

$$Q = C_u A \sqrt{2gh} \Rightarrow A = Q/C_u\sqrt{2gh}$$

Where Q = total flow (0.417 s^{-1})
   C_u = discharge coefficient (0.6)
   A = gate area $\pi D^2/4$ (m^2)
   g = acceleration due to gravity (m s^{-2})
   h = depth over gate (0.50 m)

Thus $$A = 0.417/0.6\sqrt{2 \times 9.81 \times 0.5} = 0.22 \text{ m}^2$$

and the required gate area was 0.22 m^2, which could be provided either by a circular opening of diameter 0.52 m (commercial size of 24 inches) or by a square opening with sides of 0.47 m (0.5 m) length.

The entrance gate for settled water had to be large enough to provide a flow of 0.21 m^3 s^{-1}. If the filter could be washed at 1.0 m^3 m^{-2} per min, the flow would be 0.513 m^3 s^{-1}. The following expression can be used to calculate the required area for a large square orifice with approaching velocity V_o 0.698 m s^{-1}:

$$Q = (2/3)ml\sqrt{2g}\left[\left[h_2 + \left(V_o^2/2g\right)\right]^{3/2} - \left[h_1 + \left(V_o^2/2g\right)\right]^{3/2}\right]$$

Where Q = flow (0.513 m^3 s^{-1})
   m = coefficient (0.60)
   l = side of square (1.05–h_1 metres)
   g = acceleration due to gravity ($\pm$ 9.8066 m s^{-2})
   h_1 = head on upper end (m)
   h_2 = head on lower end (1.05 m)
   V_o = approaching velocity of 0.513/(1.05 × 0.70) = 0.698 m s^{-1}

Values can be substituted and the equation rearranged to provide an expression for h_1 as shown below, which is satisfied approximately by $h_1 = \pm 0.60$ m.

$$0.513 = (2/3)0.60(1.05 - h_1)\sqrt{2 \times 9.81} \times$$

$$\left[\left[1.05 + \left[0.698^2/(2 \times 9.81)\right]\right]^{3/2} - \left[h_1 + \left[0.698^2/(2 \times 9.81)\right]\right]^{3/2}\right]$$

or

$$\left(1.05 - h_1\right)\left[1.1143 - \left(h_1 + 0.0248\right)3/2\right] = 0.2896$$

$$h_1 = \pm 0.60$$

Appropriate sides of the orifice were therefore at least 0.45 m (1.05–0.60 m) in length (equivalent to a commercial size of 18 in × 18 in). Because filter and wash rates may have been increased later, the size actually selected was 0.50 × 0.50 m, which was the same as the entrance to the main drain.

This calculation is rather cumbersome, but a similar result may be obtained by manipulating the classical hydraulic formula to give an expression for the area of the gate A in m² and the water depth in the middle of the gate h:

$$Q = KA\sqrt{2gh}$$

For $Q = 0.513$ m³ s⁻¹ and coefficient $K = 0.6$:

$$A\sqrt{h} = 0.513\Big/\left(0.6\sqrt{2 \times 9.81}\right) = 0.1931$$

The solution is to try values for A and determine h using the commercial sizes of gates. For example with a gate of 0.40 m sides, $h = 0.85$ m (because 1.05 − 0.40/2 = 0.85). Applying these values gave:

$$A\sqrt{h} = 0.4^2\sqrt{0.85} = 0.1475 < 0.1931$$

which was too small. For a gate with 0.50 m sides:

$$A\sqrt{h} = 0.5^2 = 0.2236 > 0.1931$$

which was satisfactory (and consistent with previous calculations).

The average design flow of settled water through each gate into the filter box was 0.21 m³ s⁻¹ (12 gates, total flow 2.50 m³ s⁻¹) but the gates had to handle the maximum flow, which might have been 1.5 times the average. Because even higher filtration rates are always possible, in this example 2.5 times the average flow was allowed for (0.521 m³ s⁻¹ per gate). The hydraulic expression to be used depends on whether the gate is totally or partially submerged.

The formula for a totally submerged gate is shown below, where notation is the same as used above, with V_1 and V_2 the average approaching velocities upstream and below the gate:

$$Q = KA\left[V_2 + \sqrt{2gh + V_1^2 - V_2^2}\right]$$

For a partially submerged gate the more complex formula is shown below, where coefficients K_1 and K_2 are 0.60, b is width, and h_1, h_2 and h_3 are respectively the water heights (illustrated in Figure 9.7) at the gate inlet, outlet and downstream:

$$Q = K_1 b\left(h_2 - h_3\right)\sqrt{2gh_3} + 2/3K_2 b\sqrt{2g}\left(h_3^{3/2} - h_1^{3/2}\right)$$

Partially submerged orifice

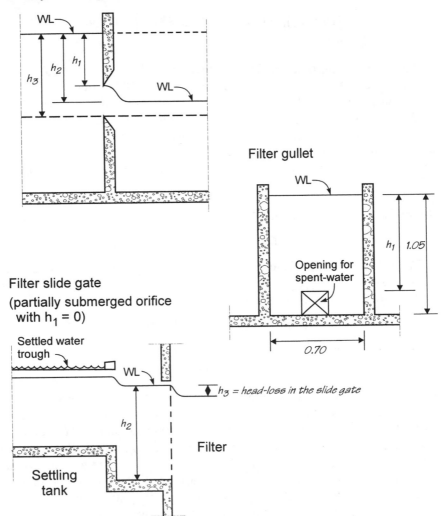

Figure 9.7 Important hydraulic characteristics, with notation, of a partially submerged orifice; a filter gullet, with dimensions for the plant under discussion; and a filter slide gate, which is a partially submerged orifice with $h_1 = 0$

It is common to use slide gates for this flow control. Assuming a head loss $h_3 = 0.05$ m and $h_1 = 0$ and substituting in the equation for a partially submerged gate:

$$\text{Flow} = 2.5(2.5/12) = 0.521 \text{ m}^3 \text{ s}^{-1}$$

$$0.521 = 0.6b(h_2 - 0.05)\sqrt{2g \times 0.05} + 2/3 \times 0.6 \times b \times \sqrt{2g}\left(0.05^{3/2} - 0.0^{3/2}\right)$$

reducing to $\qquad\qquad b + 30b(h_2 - 0.05) = 263056$

For variable filter rate operation, the deepest possible gate is the best option for inlet to the filters, and therefore the gate chosen had the same depth as the settled water collecting channel. Using this depth, the value of b was found. For the existing channel after reducing the freeboard, water height was 1.2 m which gave $b = 0.74$ m. The existing channel was only 0.70 m wide and this was the maximum that could be used. The result would be a slightly higher head loss through the gate.

Filter bed disruption is commonly caused by rapid inflow of settled water after terminating the backwash, and therefore the high kinetic energy has to be dissipated without disturbance of the filter surface. This can be achieved by directing inflow downwards into the central drain gullet. Water spills over into the filter bed once the gullet has filled, but with little risk of surface disturbance because the water level in the filter is at the gullet lip. The energy is proportional to the square of the velocity (as $m\,v^2 / 2$), encouraging large openings or large entrance gates.

In this plant, the filter bottoms had to be changed completely because the old bottom was in poor condition and could not handle the upgraded flow. A system of precast concrete corrugated bottoms was used, which could be cast on site and placed in the structure that had been prepared to receive the channels. These are shown in Figures 7.3, 9.8 and 9.9. As indicated, these triangular channels can be designed for both air and water wash. In this case, the sweeps were kept as an auxiliary washing system; the bottoms were designed only for water wash.

The central channel as shown in Figure 9.9 had a constant width of 0.60 m and a variable depth of 0.60 m at the entrance and 0.25 m at the end. The length of the precast sections was 7.93 m, being $8.20 - 2\,(0.15 - 0.015\,\sqrt{2})$ and their width was 3.49 m, being $3.75 - 2\,(0.15 - 0.015\,\sqrt{2})$.

Backwashing aims to remove deposited floc and turbidity particles and its efficiency relies mainly on the drag caused by flow over grains of sand or coal. This drag force is expressed in terms of shear, varying with grain size and specific gravity, viscosity of the washwater and expansion of the filter media. In practice, shear force is translated into velocity of backwash water and 0.8–1.0 m per min was a sufficient rate to clean the filter, whilst maintaining layering of the dual media with little mixing of sand and coal. These rates were adequate for sand of 0.65 mm effective size at 14–21 °C. Coal is much lighter than sand and causes no specific problem.

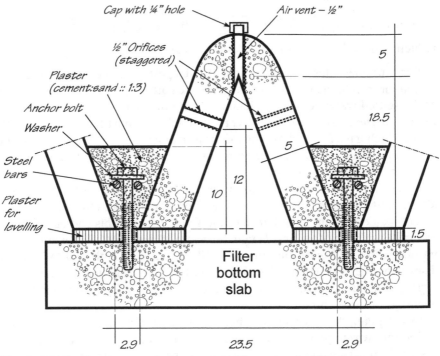

Figure 9.8 Section across a single element of the corrugated filter bottom

Dimensions of the precast triangular filter bottom drains are summarised below. For a backwash rate of 1.0 m per min which means a total flow of 0.513 m³ s⁻¹ through each filter bed:

- The number of drains was 30, each of 0.235 m width, separated by 0.03 m spaces (Figure 9.8).
- The flow in each drain was 0.0171 m³ s⁻¹ (0.513 / 30), and therefore the flow was 0.0086 m³ s⁻¹ in each side.
- The cross-sectional area of each drain was 0.0122 m².
- The flow velocity in each drain was 0.70 m s⁻¹.
- The angle of base with the horizontal was 70°.
- The thickness of drain walls was 0.05 m and of base was 0.053 m (0.05/sin 70°).
- The dimension of base was about 0.129 m (0.235 – 2 × 0.053).
- The small base was 0.03 m, height 0.133 m.

The triangular section height was 0.177 m (0.129 / (2 tan 20°)). The base and height of the triangular drain were each 0.235 m, which gave a flow area of 0.0119 m², i.e. (0.1286 × 0.185)/2.

Once the size of the drain had been determined, it was necessary to calculate the size and distribution of orifices along the sides. Distribution of

Longitudinal section

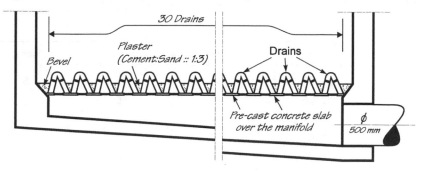

Transverse section

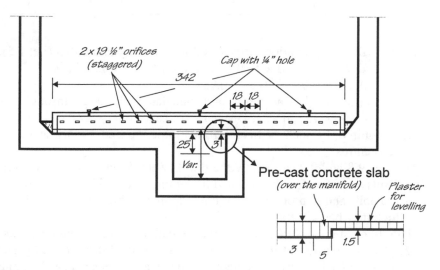

Figure 9.9 Longitudinal and transverse sections of the pre-cast concrete corrugated bottoms

backwash water has to be as uniform as possible and this is achieved by having quite a high head loss in each orifice. Water seeks the easiest route but with high head losses in each orifice there was no easy exit and distribution was good.

Orifices were constructed using a minimum standard PVC pipe of 0.0127 m diameter, such that the area of each orifice was 0.0001267 m^2, and the orifice length (thickness of the drain walls) was 0.05 m. Orifice head loss was 1.5 m and discharge through each was calculated by:

$$Q = C_d A \sqrt{2gh} \text{ where } C_d = 1 \Big/ \sqrt{\Sigma C_1 + \Sigma f(L/d)}$$

Table 9.2   Profile (size, depth) of support gravel in filter beds

	Gravel size		Depth	
	mm	inches	cm	inches
Top	25–50	1–2	12.5	5
	12–25	$^1/_2$–1	5	2
	6–12	$^1/_4$–$^1/_2$	5	2
	3–6	$^1/_8$–$^1/_4$	10	4
	6–12	$^1/_4$–$^1/_2$	5	2
	12–25	$^1/_2$–1	5	2
Bottom	25–50	1–2	10	3

Where ΣC_1 = summation of localised losses, entry and outlet (i.e. 1.5).
$\Sigma f(L/d)$ = loss due to flow in tube, where $f \approx 0.237$ (i.e.
0.237 × 0.05 / 0.0127 = 0.933).

The coefficient C_d was 0.641 and calculated discharge through each orifice was 0.0004405 m^3 s^{-1}. Discharge in each drain was 0.0171 m^3 s^{-1}, which required 19 such orifices on each side, spaced evenly at 0.18 m between centres.

Figures 9.8 and 9.9 show how the drains were fixed to the filter bottom slab. Bolts were fixed in the slab and the drains were held in place by grout over the bolts and reinforcing steel. Each drain had three vents for relieving air which may become trapped.

The filter support bed was reverse gradation gravel, occupying about 0.525 m in seven layers (Table 9.2). Sand from existing filters of effective size c. 0.65 mm and uniformity coefficient 1.15 was washed, cleaned, and regraded to make the 25 cm sand layer. The coal layer was 45 cm deep with an effective size of 1.20 mm and uniformity of 1.25. These specifications were based on long experience with dual media filter beds. Sizes were selected for less developed regions where resources for sophisticated filter operation are not always available. The media have excellent storage capacity which provides good floc penetration for long filter runs, whilst maintaining filtered water quality such that turbidity is usually less than 0.50 NTU. A detailed description of the method of placing media in the filter box is given in Section 4.5.

These filters were operated in the declining rate mode of control, using a perforated disc in the filter outlet line which limited flow in the clean filter to a calculated maximum determined by plant design capacity. In these filters the average rate needed to pass the upgraded volume was 12.36 m^3 m^{-2} per hour. Directly after backwash they filtered at 18.54 m per hour

(0.0052 m^3 s^{-1}, the limit imposed by the perforated disc) declining to 6.18 m per hour during the filter run. The area of the disc orifices was calculated as follows:

$$Q = C_d A \sqrt{2gh} \Rightarrow A = 0.0052 / 15 \times 5.86 = 0.0006 \text{ m}^2$$

Where Q = discharge to be allowed (0.0052 s^{-1})
          g = acceleration due to gravity, 9.81 m s^{-2}
          h = head to media top (1.75 m)
          C_d = discharge coefficient (1.5 in this case)

The required area of 0.00059 m^2 could have been provided by five holes formed with a standard PVC pipe of 1.27 cm diameter, or perhaps by nine holes lined with $^3/_8$ inch pipe. An alternative solution was to install a disc with adjustable orifices as shown in Figure 8.23.

Chapter 10

OPTIMISATION AND UPGRADING OF A PLANT FROM 20 l s^{-1} TO 50 l s^{-1}

10.1 Assessment of original plant

This was a typical small plant built about 30 years ago. Raw water was taken from a stream that runs through a hilly wooded area. Although flow can vary rapidly, heavy vegetation reduced the chance of high turbidity. Raw water entered the plant by gravity from an intake 2 km upstream.

The original design provided for 20 l s^{-1}, allowing 25 per cent overload (25 l s^{-1} maximum) but the town served by this plant had grown such that more water was required to satisfy demand. The original structures shown in Figure 10.1 are described below.

10.1.1 Gravity pipeline and inlet chamber

The pipeline from the intake was 200 mm in diameter and had capacity for only 25–30 l s^{-1}. The inlet chamber was a square concrete box 1.10 m^2 in area and 2.50 m deep. Raw water entered this box through a 90° upturned elbow and flowed through a Parshall flume, where flow was measured by a meter rod attached to the wall and calibrated in litres per second. The coagulant (alum) was applied at about 20 per cent concentration in a single stream at the entrance box.

10.1.2 Flocculation

From the flume, water flowed to the hydraulic flocculator, which was 2.00 m wide, 9.60 m long and 1.90 m deep. Velocity at the bends of the 19 baffles varied from 0.45 m s^{-1} down to 0.20 m s^{-1}. The water flowed to the settling basin through a short channel 0.50 m wide and 0.80 m deep. The freeboard was 0.30 m.

10.1.3 Settling basins

Two settling tanks received flocculated water. Each was 3.50 m wide, 14.00 m long, and the depth varied from 3.50 m at the inlet to 4.00 m at the sludge drain to 3.00 m at the basin outlet. The entrance to each basin was through a gate 0.25 m wide and 0.50 m deep, into a rectangular box across the section with large round ports in the bottom (see Figure 9.6).

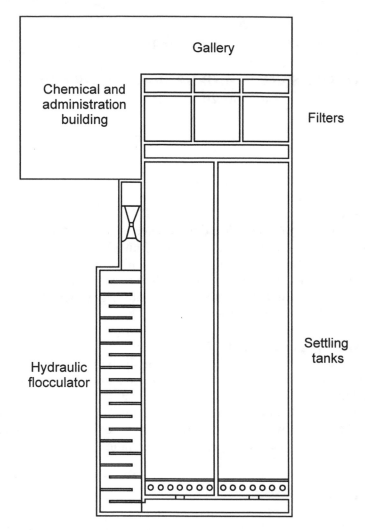

Figure 10.1 Plan view of existing plant, design capacity 20 l s^{-1}

Holding time in the basins was 4 hours. The settled water was removed by a weir across the outlet end of the basin. The settled water collection channel was 0.69 m wide and 0.80 m deep, conducting water to the filters through circular sluice gates 0.20 m in diameter.

10.1.4 Filters
There were three filters, each 2.20 × 2.25 m, with sand beds 0.70 m thick and a support gravel bed of 0.45 m. The underdrain system was based on a 0.25 m square reinforced concrete central manifold. Six asbestos cement laterals on

Table 10.1 Characteristics of raw water during the rainy and dry seasons for a treatment plant with an original capacity of 20 I s^{-1}

	Rainy season			Dry season		
	Max.	Min.	Average	Max.	Min.	Average
Turbidity (NTU)	100	30	55	25	15	18
Colour (TCU)	47	25	35	18	7	12
Alkalinity (mg I^{-1})			15			21
pH			5.8			6.2
Temperature (°C)			10			28
Bacterial contamination	Very low					

each side were 75 mm in diameter with perforations for the entrance of filtered water and distribution of backwash water. There were no wash water troughs. A 0.70 × 1.00 m gullet at the end of the filter was used to collect backwash water.

The backwash was provided from an elevated tank through a 200 mm cast iron pipeline. This tank had no anti-vortex baffles, which meant that air was entrained in the backwash water during the latter part of the wash.

Each filter had a rate of flow controller which did not operate properly. The filter rate was conventional at about 5.00 m^3 m^{-2} per min.

10.1.5 Chemical building
The three-storey chemical building provided storage space for chemicals, feeders, day tanks for preparing coagulant, lime-slaking facilities and respective feeders, chlorine cylinders and chlorinators. The elevated storage tank for backwashing was situated on top of this building.

Operators' facilities were also housed in the chemical building (laboratory, office, kitchen, bath and dressing room, and a dining room) and each of these facilities was rather small.

10.1.6 Raw water
The characteristics of raw water differed between the rainy and dry seasons (Table 10.1).

10.2 Plant performance
Design contributed to the poor performance of this plant, but its operation was also unsatisfactory. The operators had only a limited basic knowledge. They had received no special training in plant operation, and did not

appreciate the importance of quality control and laboratory tests to obtain optimisation. Moreover, there was little laboratory equipment and none of it functioned well. The town itself was small and had limited resources for paying operators and for training.

Alum was applied undiluted from the day tank at about 20 per cent concentration and flowed into the raw water as a single stream above the Parshall flume. The result was uneven dosing of the raw water. A small proportion received too much alum, whilst most of it was underdosed.

Flocculation was not complete because the system provided mixing (with energy input through a high velocity gradient) only around the ends of the baffles. There was almost no turbulence, and therefore little mixing along their length. Samples collected before or after the unit, and then flocculated in the laboratory, were more superior in floc formation than samples taken from the outlet with no further treatment. This was clear evidence of deficiency in the existing flocculation unit.

The settling basins were likely to perform poorly because of incomplete initial dispersion of alum and incomplete flocculation, but the design of the basins themselves could also have been improved. The entrance structure allowed short-circuiting currents to form, so that there was insufficient time for floc to settle from a proportion of the water. The short weir at the end of the basin had a very high overflow velocity, creating an upsweeping current that carried over a great deal of excess floc. Both of these problems contributed further to the turbidity of the settled water, which was always too high, in the range 12–20 NTU.

Excessively high settled water turbidity had caused general deterioration in the filters, which showed cracks and had a high percentage of mud balls. The backwash was uneven across the filter area, which suggested some breakage in the filter bottoms. Filtered water turbidity was usually above the WHO guideline value of 5 NTU and often in the range 10–15 NTU.

Alum preparation in the day tanks was a routine procedure and operators knew how much alum to put into each tank of water, consistently producing a concentration of 20 per cent. The process thereafter was erratic because the feeders were operating poorly. For much of the time the raw water was underdosed and sometimes it was overdosed, at cost to both quality and economy.

10.3   Upgrading parameters determined by jar testing

Both alum and ferric chloride performed well. Alum was selected because the plant operators were more familiar with it and the supply and handling was easier. A high velocity gradient of about $1000 \ s^{-1}$ was required for mixing the coagulant with raw water. The coagulant needed to be applied as a more dilute solution, having an alum concentration of no more than 0.5 per cent.

During the wet season 14–16 mg l^{-1} alum and 3–3.5 mg l^{-1} of lime were needed. Direct filtration was possible during the dry season, using 1.5–2.0 mg l^{-1} alum for destabilisation.

The optimum flocculation time during the wet season was 25 minutes applying tapered energy input of 5 minutes with velocity gradient 60–70 s^{-1}, 7 minutes at 35–45 s^{-1}, then 13 minutes at 15–20 s^{-1}. In the dry season, 5–7 minutes of mixing at 60–70 s^{-1} was satisfactory for direct filtration. The times for flocculation and settling were the same as for the wet season.

The surface loading required to remove 90–95 per cent of turbidity in the dry season was 2.5 cm per min, under conventional treatment. In the wet season, a loading of 3.2 cm per min achieved such removal.

Adjustment of the pH would always be necessary for full flocculation-based pretreatment, but was not required if destabilisation and direct filtration were being used in the dry season.

The possibility of treatment by direct filtration under dry season conditions was tested using Whatman No. 40 filter paper after dosage with alum. A dose of 1.5–2.0 mg l^{-1} consistently provided filtered water turbidity of less than 1.0 NTU.

10.4   Design of modifications for upgrading and optimisation
The modifications were designed to provide a treatment capacity of 50 l s^{-1}, to improve treated water quality, to keep the upgrading costs as low as possible, to provide the plant with means of easily controlling the treatment process and to reduce the cost of treatment per cubic metre.

10.4.1 Initial dispersion of coagulant
The existing flume was too small to measure 50 l s^{-1} but could accurately measure 25 l s^{-1}, therefore the raw water was divided by a splitter box with two sharp-crested weirs, each 0.40 m long.

The channel from one weir went to the Parshall flume for flow measurement and the other led to the new flocculation basin, which was parallel and adjacent to the existing basin. Coagulant was applied at the new weirs in the splitter box for good dispersion. The fall at the weir was a little over 0.10 m to ensure a velocity gradient of more than 1,000 s^{-1}.

10.4.2 Flocculation
The new flocculation basin was designed to provide half of the water with the velocity gradients indicated by jar testing. The existing basin was modified to provide the same energy input to the remainder of the water. Another modification was the installation of fins vertical to the baffles as shown in Figures 8.11 and 8.12. These fins induced turbulence along the length of the baffle to

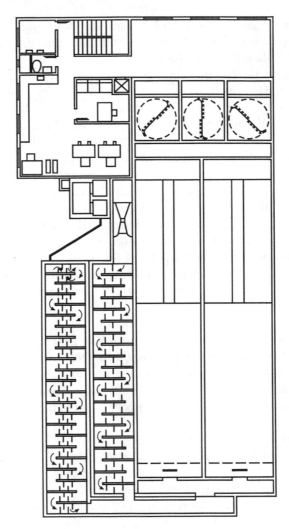

Figure 10.2 Layout of optimised and upgraded treatment plant, from 20 l s^{-1} to 50 l s^{-1}

increase mixing efficiency, thereby reducing the difference between the high velocity gradient at the bends and the very low gradient along the baffle.

The basins both provided 25–26 minutes of flocculation time. Each basin had 19 baffles (Figure 10.2). By varying the size of the openings at the baffle ends, velocity could be set by design to provide the desired steps of velocity gradient: 0.15 m s^{-1} at the first three baffles (giving 65 s^{-1}), 0.105 m s^{-1} at the next three baffles (for 40 s^{-1}) then 0.065 m s^{-1} at the remainder (20 s^{-1}).

10.4.3 Settling basins

The settling basins were designed to remove floc of a certain settling velocity as determined in the jar testing, which in this case was 2.5 cm per min (36 m^3 m^{-2} per day).

Excessive detention time has no direct value to the clarification process. Heavy, dense flocs will settle in minutes but light, poorly formed flocs may remain suspended for hours because of thermal, wind-induced or density currents. Basin loading therefore depends on formation of fast-settling floc through proper dispersion of dilute coagulant and good flocculation.

Surface loading is the most important parameter for good settling, but mean horizontal velocity must also be limited to avoid scouring and carriage of deposited flocs along the basin. For alum sludge in conventional rectangular settling basins, the proportion of horizontal velocity to settling velocity is $(8k/f)^{1/2}$ where k is a factor measuring the shape of particles and f is the friction coefficient of the Weisbach–Darcy formula. This value lies in the range 8–18, and because the total surface area is 98 m^2 (two settling basins, each 3.50 × 14.00 m), the acceptable horizontal flow is between 0.20 m per min and 0.45 m per min. With a proposed flow of 50 l s^{-1} the expected mean horizontal velocity through the basins of the present section was 0.136 m per min. This means that with some sludge accumulation, the available cross-sectional area was sufficient to accommodate the increased flow within desired 0.20–0.45 m per min limits.

Horizontal flow settling plates were installed in the final 5.10 m of the basin to increase surface area (to 120 m^2), thereby improving the clarification. Between the settling plates and the outlet a perforated baffle with a high head loss (about 2 cm) was installed, for an even distribution of flow through the plates. A system of settling plates, with the appropriate dimensions and layout, is shown in Figures 8.15–8.18.

The inlet portion of the basin continued to provide plain settling of the heavier floc while the lighter particles were removed in the settling plates.

Figure 10.2 shows changes in the outlet of the old flocculation basin and the inlets to the settling basins. Both flocculation basins emptied into a new channel which fed the settling basins. The two entrance ports had to be enlarged for passage of the new flow without causing turbulence and floc breakage. These new gates required a section of 0.50 m^2 each.

10.4.4 Filters

The three filters provided an area of 14.85 m^2 (3.00 × 2.25 × 2.20 m) and the filter rate under the existing conditions was 117–150 m per day (5–6 m per hour) depending on the overload. For the new flow, the rate with the existing filters was 291 m per day (12.12 m per hour), which is more than double the original rate but well within the possibilities of the plant.

Good filters with effectively pretreated water can be loaded beyond conventional practice with satisfactory results. Many plants in the USA operate at 3–5 times the usual 5 m per hour rate, yet perform very well. All such heavily loaded filters use dual anthracite and sand media. The filters in this plant comprised a 0.45 m layer of local coal over 0.25 m of sand, with about 0.45 m of support gravel laid in a reverse gradation of size. Suitable sections and areas with dimensions are as shown in Figure 8.20.

The filter control was by declining rate with all three filters controlled from one flow control box. Such an arrangement is shown, with dimensions, in Figures 8.21 and 8.22. This system required the removal of the existing rate controllers.

The concept of declining rate may be new to some engineers but it is not very complex. The maximum filtration rate of a clean filter is limited by the level of the weir in the control box, along with the head loss in the filter, and by piping with a fixed available head and a fixed head loss. The filter must operate within these limits. Immediately after washing, the filter starts at its maximum rate. As clogging begins the rate declines, because flow is obstructed by deposition of floc in the pores of the filter media. The filtering rate declines continuously until it has used all the available head, causing the water to overflow. Obviously, the operator should not allow this final condition — the filters should be washed at regular intervals when the rates have declined to a point where their sum is less than the amount to be treated.

Ideally, experiments with pilot filters should be done to determine the required filterbed for the available head and the water quality desired for public supply. When no such studies have been made, backwashing is done whenever either the available head or the water quality criteria are surpassed. Filters may typically require washing after every 24, 36, or 40 hours of filter run and experience quickly shows the appropriate washing pattern. The three filter beds in this plant needed to be washed to a schedule which had filters at different stages of the filter cycle: one recently washed, one in mid-cycle, and one nearing the end of its run. Once this pattern was set, operators knew exactly when to wash each filter (subject to stable quality of the pretreated water).

The filters needed to be completely rebuilt. Because of the health risks associated with asbestos, the old asbestos cement laterals were no longer accepted for potable water, and had to be substituted with PVC pipes. The diameter of the laterals was 125 mm to provide a high-rate backwash of up to $1.00 \text{ m}^3 \text{ m}^{-2}$ per min. Details and dimensions of this type of filter bottom with the central manifold and laterals are shown in Figures 9.8 and 9.9.

Each filter was supplied with a rotary sweep with jet-propelled arms to provide rotation. This auxiliary cleaning during backwash was necessary to break up large pieces of floc, which collected in the coal and formed mud balls too quickly without some auxiliary agitation.

The piping of the filters also had to be changed to accommodate high filter and backwash rates, as discussed in Chapter 8 (Figures 8.21 and 8.22 show the layout of piping).

10.4.5 Chemical building and handling

Two new alum preparation tanks had to be built on the top floor of the chemical building so that the alum could be fed by gravity. Each tank had a capacity of 1,000 litres of 10 per cent alum solution (100 kg of dry alum added to 900 kg of water). An allowance was needed if there were insolubles in the dry alum, for example 105 kg of dry alum with 5 per cent insolubles. The alum solution flowed down to a constant head, float-controlled alum feeder which could be adjusted to give an alum dose within the anticipated range.

Jar tests suggested a maximum dose of 16 mg l^{-1} but for response to future problems it was wise to allow for 25 mg l^{-1}. The design raw water flow was 3,000 litres per min (50 l s^{-1}) and therefore 75,000 mg per min may be required. Because the 10 per cent alum solution in the day tanks contained 100 mg ml^{-1}, the maximum flow dispensed by the feeder was 750 ml per min. Tests for treatment by direct filtration during the dry season suggested 1.5–2.0 mg l^{-1} alum, and therefore the feeder also needed to be able to dispense as little as 45–60 ml per min with accuracy.

It is critical to the whole of the treatment process that accurate dosing is maintained. There were two feeders for alum and two for lime, providing a standby in case of failure. To fulfil its role, the standby system had to be provided with a high standard of maintenance.

Until recently the importance of initial dispersion of the coagulant in the raw water was not well understood. It is now clear that the aqueous chemistry of coagulant ions is complex, both aluminium and ferric ions undergoing a series of reactions with the hydrosoil forming polymeric species in just a fraction of a second. Ideally, all the coagulant should be dispersed into all the raw water instantly. This is not physically possible with large quantities but there are means to achieve a very rapid dispersion.

The raw water was spread over a long weir where the nappe over the weir was less than 10 cm depth. Through a diffuser as shown in Figures 8.8 and 8.9, very dilute coagulant solution was applied to this shallow, highly turbulent flow. Coagulant was applied in the most dilute state that did not react with dilution water, i.e. 0.20–0.30 per cent concentration or 0.50 per cent to be cautious. This system produced the most favourable conditions for dispersing all the coagulant in all the water within a fraction of a second.

Immediately prior to application through the diffuser at the weir, the alum solution was diluted from 10 per cent down to 0.5 per cent. This required a maximum of 15 litres per min of dilution water, with flow controlled by a box within a V-notch weir. Calibrations on the side of the weir allowed the operator to adjust the amount of dilution water to the dose being applied. Figures 8.8–8.10 show the chemical preparation, feed dilution and diffuser dosing.

Appendix I

SOURCES OF FURTHER INFORMATION

APHA 1998 *Standard Methods of the Examination of Water and Wastewater.* 20th Edition, American Public Health Association, Washington DC.

ASCE 1969–1972 Proceedings published in *Journal of the Sanitary Engineering Division* and *Journal of the Environmental Engineering Division.* American Society of Civil Engineers, New York.

AWWA 1974 *Upgrading Existing Water Treatment Plants.* American Water Works Association Seminar Proceedings, Annual Conference, Boston, MS, June 15–16, 1974.

AWWA (American Water Works Association) 1990 *Water Quality and Treatment — A Handbook of Community Water Supplies.* F.W. Pontius [Ed.] 4th Edition, McGraw-Hill, New York.

AWWA 1991 *Criteria for the Renovation or Replacement of Water Treatment Plants.* American Water Works Association Research Foundation, Denver, CO.

AWWA 1992 *Operational Control of Coagulation and Flocculation Processes.* Manual of Water Supply Practices No. M37, American Water Works Association, Denver, CO.

AWWA 1997 *Simplified Procedures for Water Examination (Manual M12).* American Water Works Association, Denver, company

AWWA (American Water Works Association) and ASCE (American Society of Civil Engineers) 1998 *Water Treatment Plant Design.* 3rd Edition, McGraw-Hill, New York, 806 pp.

Babbitt, H.E, Doland, J.J. and Cleasby, J.L. 1962 *Water Supply Engineering.* 6th Edition, McGraw-Hill series in Sanitary Engineering and Science, McGraw-Hill, New York, 672 pp.

Brikké, F. 1993 *Management of Operation and Maintenance in Rural Drinking Water Supply and Sanitation: A Resource Training Package.* World Health Organization, Geneva.

Chadwick, A. and Morfett, J. 1993 *Hydraulics in Civil Engineering.* 2nd Edition, E&FN Spon, London.

Cheremisinoff, N.P. and Cheremisinoff, P.N. 1993 *Water Treatment and Waste Recovery—Advanced Technology and Application.* Prentice Hall Series in Process Pollution and Control Equipment, PTR Prentice Hall, New Jersey.

Craum, G.F. [Ed.] 1993 *Safety of Water Disinfection: Balancing Chemical & Microbial Risks*. Ilsi Press, Washington, D.C.

Culp, J.L. and Culp, R.L. 1974 *New Concepts in Water Purification*. Van Nostrand Reinhold Environmental Series, New York.

EPA 1990 *Technologies for Upgrading Existing or Designing New Water Treatment Plant Facilities*. Technology Transfer Manual No. EPA/625/4-89, Environmental Protection Agency, Cincinnati, OH.

EPA 1991 Optimizing Water Treatment Plant Performance Using the Composite Correction Programme. Technology Transfer Manual No. EPA/625/6-91/027, Environmental Protection Agency, Cincinnati, OH.

Fair, G.M., Geyer, J.C. and Okun, D.A. 1966 *Water and Wastewater Engineering. Volume 1 Water Supply and Wastewater Removal*. Wiley, New York.

Fair, G.M., Geyer, J.C. and Okun, D.A. 1968 *Water and Wastewater Engineering. Volume 2 Water Purification and Wastewater Treatment and Disposal*. Wiley, New York.

Hall, T. [Ed.] 1997 *Water Treatment Processes and Practices*. 2nd Edition, Water Research Centre, Swindon.

Hudson, Jr., H.E. 1981 *Water Clarification Processes — Practical Design and Evaluation*. Van Nostrand Reinhold Environmental Engineering Series, Van Nostrand Reinhold Company, New York.

Hutton, L.G. 1983 *Field Testing of Water in Developing Countries*. Water Research Centre, Swindon.

James M. Montgomery, Consulting Engineers Inc. 1985 *Water Treatment Principles and Design*. John Wiley & Sons, New York.

Jordan, J.K. 1990 *Maintenance Management*. American Water Works Association, Denver, CO.

Kawamura, S. 2000 *Integrated Design and Operation of Water Treatment Plants*. 2nd Edition, John Wiley & Sons, Inc, New York.

Kay, M. 1998 *Practical Hydraulics*. E&FN Spon, London.

Kerri, K.D. 1994 *Water Treatment Plant Operation — A Field Study Training Program*. 3rd Edition (2 volumes), California State University, Sacramento.

Kerri, K.D. 1999 *Small Water System Operation and Maintenance — A Field Study Training Program*. 4th Edition, California State University, Sacramento.

Letterman, R.D. 1991 *Filtration Strategies to Meet Surface Water Treatment Rule*. American Water Works Association, Denver, CO.

Mallevialle, J. *et al*. [Eds] 1992 *Influence and Removal of Organics in Drinking Water*. Lewis Publishers.

McNown, J.S. 1954 Mechanics of manifold flow. Paper No. 2714, *Transactions of the American Society of Civil Engineers*, Vol. **119**.

Najm, I.N. *et al.* 1991 *Control of Organic Compounds with Powered Activated Carbon.* Subject Area: Water Treatment and Operation, American Water Works Association Research Foundation, Denver, CO.

Schulz, C. and Okun, D. 1984 *Surface Water Treatment for Communities in Developing Countries.* Intermediate Technology Publications, London.

Stenquist and Kaufman SERL Report-72-2, University of California, Berkeley.

Summers, S.S. *et al.* 1992 *Standardized Protocol for the Evaluation of GAC.* AWWA Research Foundation, Subject Area: Water Treatment, American Water Works Association, Denver, CO.

Tebbutt, T.H.Y. 1998 *Principles of Water Quality Control.* 5th Edition, Butterworth Heinemann, Oxford.

Twort, A.C., Ratnayaka, D.D. and Brandt, M.J. 2000 *Water Supply.* 5th Edition, Arnold, London.

van Duuren, F.A. [Ed.] 1997 *Water Purification Works Design.* Water Research Centre, Pretoria.

Vrale and Jordan 1971 Initial mixing of coagulant in raw water. *Journal of AWWA*, 63–52.

Weber, Jr., W.J. 1972 *Physicochemical Processes for Water Quality Control.* Wiley-Interscience, John Wiley & Sons, New York.

White, G.C. 1999 *Handbook of Chlorination and Alternative Disinfectants.* 4th Edition, John Wiley, New York, 1569 pp.

WHO 1993 *Guidelines for Drinking Water Quality. Volume 1 Recommendations.* 2nd Edition (plus addendum, 1998), World Health Organization, Geneva.

WHO 1994 *Operation and Maintenance of Urban Water Supply and Sanitation Systems — A Guide for Managers.* World Health Organization, Geneva.

WHO 1996 *Guidelines for Drinking Water Quality. Volume 2 Health Criteria and Other Supporting Information.* 2nd Edition, World Health Organization, Geneva.

WHO 1997 *Guidelines for Drinking Water Quality. Volume 3 Surveillance and Control of Community Supplies.* 2nd Edition, World Health Organization, Geneva.

WHO 2000 *Tools for Assessing the O&M Status of Water Supply and Sanitation in Developing Countries.* World Health Organization, Geneva, In press.

Wyatt, A. 1989 *Guidelines for Maintenance Management in Water and Sanitation Utilities in Developing Countries.* US Agency for International Development (WASH), Washington DC.

Yao, K.M. 1972 Hydraulic control for flow distribution. *Journal of the Sanitary Engineering Division, ASCE*, **98**:(SA2), 275–285.

Yao, K.M. 1970 Theoretical study of high-rate sedimentation. *Journal of the Water Pollution Control Federation*, **42**, 218.

Appendix II

GLOSSARY OF TERMS

Alum Aluminium sulphate ($Al_2(SO_4)_3$): the most widely used coagulant throughout the world.

Around the end mixing A channel where the water passes alternately around the ends of the baffles.

Blind or stilling baffle A solid wall baffle placed to absorb energy of water entering a basin.

Coagulated water Raw water after it has received destabilising chemicals that remove the electric charges from the turbidity particles and allows them to combine together and grow.

Collecting or combining channels These channels receive water from a series of basins or pipes and carry it to the next treatment unit or to waste.

Declining rate This is achieved with an adjustable plate at the filter outlet to control the maximum rate, which declines as the filter clogs and needs washing. This system is most widely used now because of its economy and simplicity.

Deep bed filter Single media of coal with a depth of 1–1.3 m, effective size of 1.5–3.5 mm and highly uniform (about 1.2 uniformity). Very effective with direct filtration.

Direct filtration Applying water to the filter directly after coagulation with little or no flocculation and no settling. The filter does the work. Very economical approach where feasible.

Disinfected water Filtered water having received a disinfecting chemical, usually chlorine, and held in the acid range for a period of 30 minutes in a baffled basin to control short-circuiting.

Distribution channels These distribute the water form a channel or header to a series of basins or pipes.

False filter bottoms with nozzles Plastic nozzles of calculated size and number.

Ferric chloride ($FeCl_3$) A coagulant that is effective over a wide range of pH. It produces a heavier floc than alum.

Filter media Sand is the most commonly used media in the less developed countries. The depth is usually 60–70 cm and the effective size ranges from 0.6 to 0.8, but both depth and size can vary widely. Dual media, of materials of different specific weight, is used throughout the industrialised countries; it usually consists of a layer of anthracite coal over sand. The combination can take a much higher load — four or more times the old conventional load.

Filter support   The filter media is supported by a layer of round stones, the sizes of which vary from approximately 2–3 mm to 25–50 mm, with the large stones on the bottom and the small stones on top.

Filtered water   Settled water that has passed through a porous granulated media at a controlled rate. The media can be sand or coal or a combination of the two.

Flocculated water   The coagulated water is mixed to provide opportunities for particles to come together and form settleable floc. The mixing energy input is high at the beginning and low at the end of the flocculation process.

Gates   Openings in channels or basins to allow water to flow in or out.

Gravity   Energy applied by means of a difference in elevation.

Gutter   A channel along the filter side either to distribute settle water over the filter or to receive backwash wastewater to discharge to waste.

Hydraulic flocculation   Mixing energy applied by gravity in a baffled channel or in a basin.

Impeller mixing   Pushes water from the centre to the sides of the basin causing non-uniform distribution. Mixing is high at the ends of the paddles and low in the centre at the axis.

Leopold plastic block   A filter precast and designed for air and water back wash.

Lime   The most commonly used alkali to stabilise water before distribution. Three forms are available: limestone ($CaCO_3$), quicklime (CaO) and hydrated lime ($Ca(OH)_2$). Hydrated lime is the most commonly used form for stabilisation.

Manifold   A main channel providing equal flow distribution to lateral channels.

Mechanical energy   Energy provided through mechanically operated equipment.

Mechanical flocculation   Mixing energy applied by mechanical means.

Over and under mixing   A baffled channel where the water passes over one baffle and under the next, etc.

Perforated baffle   A baffle with a calculated number and size of perforations to introduce a head loss which serves to distribute water uniformly across the inlet or outlet of a basin.

Precast corrugated filter bottom   Can be designed for specific flow rates of water and air and can be fabricated on the construction site.

Propeller mixing   Power applied in a basin providing an axial flow pattern for good distribution.

Rate control devise   A constant rate is most commonly used in older treatment plants. The devise maintains the rate constant throughout the filter run. These devises are almost always out of adjustment or completely abandoned.

Raw water   Water that comes to the plant inlet from the source — river, lake or reservoir.

Reverse gradation support   An additional layer on top of the filter support, grading from small to large stones. This type of support is much more resistant to shocks and displacement forces.

Settled water   Flocculated water after removal of settleable flocs in appropriate basins.

Slide gate   A gate with a solid light metal or plastic piece that can slide up or down in the opening to control the flow in or out.

Soda ash (Na_2CO_3)   Used for stabilising water for distribution.

Stabilised water   Disinfected water to which alkali has been added to bring the pH to non-corrosive condition.

Trough   Structures over the filter bed either for distributing settled water in the filter or to receive backwash wastewater for discharging to the gutter.

Index